The Calculation of Atomic Collision Processes

The Calculation of Atomic Collision Processes

KENNETH SMITH
University of Nebraska

WILEY-INTERSCIENCE,
A DIVISION OF JOHN WILEY & SONS, INC.
NEW YORK · LONDON · SYDNEY · TORONTO

Library of Congress Catalog Card Number: 78-168645

ISBN 0-471-80000-7

Printed in the United States of America.

10 9 8 7 6 5 4 3 2 1

Preface

In numerous scientific and technical fields, including astrophysics, aeronomy, gas lasers, controlled thermonuclear fusion, and MHD generators, progress depends heavily on the data derived from a study of atomic and molecular collision processes. A truly useful theoretical description of these processes must be in an algorithmic form, so that it can provide *numbers*. However, most courses in collision theory follow the traditional approach of emphasizing the various theoretical manipulations of the Schrödinger equation and derive their inspiration from the classic book by Mott and Massey. Indeed, this emphasis has dominated the half-dozen monographs published on the subject during the 1960s. The purpose here is to stress some of the theoretical avenues that can be exploited by the application of digital computers, and to emphasize that models in theoretical physics must be put to quantitative test. To this end, detailed descriptions are given of some of the numerical methods that have been developed to aid in the computation of cross sections. What is emphasized here is the invention of algorithms rather than numbers as such. The computational physicist, once he has his algorithm expressed as a subroutine, can generate all the numbers he wants.

This book is aimed at students who have had an introductory course in wave mechanics. It is meant to be self-contained, and most steps in the theoretical development of the formulae are given. Consequently, the time available for a two-semester course has limited the selection of topics of physical interest and has restricted the discussion of the theory to the successful eigenfunction expansion method. Part 1 is readily understood by mixed classes of experimentalists and theorists; Part 2 is intended as the second-semester course for students who want to be able to read the current literature.

The topical emphasis is on those processes that are most likely to receive increasing attention in the next decade—differential cross sections and spin polarization for electron impact, as well as total and differential cross sections in photoabsorption. Unfortunately, time does not permit a discussion of how spin-dependent forces modify the formulae given here. However, the formulation of the atomic problem with spin-dependent forces is so close to that of low-energy nuclear physics, which is well documented, that perhaps no apology is necessary.

An important part is the solution of the problems that demonstrate explicitly to the student how the digital computer has reduced the research problem of not too long ago to a weekend exercise, when computing facilities are readily accessible. I am convinced that the use of the digital computer has added a new dimension to education and has led students to a better understanding of the material, since realistic models can be solved in a fraction of the time taken to solve idealized models in the more conventional ways of instruction. Indirectly, the computer assists in the learning process by demanding a precise and algorithmic statement of the problem, always clarifying and extending the student's knowledge.

KENNETH SMITH

Lincoln, Nebraska
May 1971

Acknowledgments

The techniques discussed in these pages would not have been developed but for the very generous support over the last decade or so, particularly in computer time, of various agencies including the Argonne National Laboratory, Lockheed Missile and Space Corporation, the British Science Research Council, the Universities of London and Nebraska, the National Science Foundation, and especially Kirtland Air Force Base.

K. S.

Contents

The Calculation of Atomic Collision Processes

1

Single-Channel Problems

1.1. POTENTIAL SCATTERING

1.1.1. Coordinate Frames

Consider particles of type A colliding with particles of type B and assume that the particles interact through a real two-particle potential $V(r_{AB})$, which depends only on the scalar distance between A and B. The classical Hamiltonian, representing the total energy, for a system composed of two such particles may be written as the sum of the kinetic energies and the potential energy, namely,

$$H = \frac{p_A^2}{2m_A} + \frac{p_B^2}{2m_B} + V(r_{AB}), \tag{1.1}$$

where m_A and m_B are the masses and $\mathbf{p}_A$ and $\mathbf{p}_B$ are the vector momenta of particles A and B in the laboratory reference frame. In the steady state, the total energy of the system E is a constant, that is, it is time-independent and the Schrödinger equation is

$$H\Psi(\mathbf{r}_A, \mathbf{r}_B) = E\Psi. \tag{1.2}$$

We define the relative coordinate $\mathbf{r}_{AB}$ as

$$\mathbf{r}_{AB} = \mathbf{r}_A - \mathbf{r}_B = \boldsymbol{\rho} \qquad \boldsymbol{\rho} = (\rho\theta\phi) \tag{1.3}$$

and the coordinate of the center-of-mass of this two-particle system in the laboratory frame, $\mathbf{R}$, by

$$M\mathbf{R} = (m_A + m_B)\mathbf{R} = m_A\mathbf{r}_A + m_B\mathbf{r}_B. \tag{1.4}$$

If the time derivatives of Eqs. 1.4 and 1.3 are taken, and the results added together [after multiplying (1.3) by m_B], as well as subtracted from each other [after multiplying (1.3) by m_A], we obtain

$$\mathbf{P}_A = \mu\dot{\boldsymbol{\rho}} + m_A\dot{\mathbf{R}}$$

and

$$\mathbf{P}_B = -\mu\dot{\boldsymbol{\rho}} + m_B\dot{\mathbf{R}}$$

where the relative, or reduced, mass of the two-particle system, μ, is defined by

$$\mu \equiv m_A m_B (m_A + m_B)^{-1}.$$

The relative momenta and the momenta of the center-of-mass are defined by

$$\mathbf{p}_\rho \equiv \mu\dot{\boldsymbol{\rho}} \qquad \text{and} \qquad \mathbf{p}_R \equiv M\dot{\mathbf{R}}.$$

With these definitions, it is a straightforward matter to show that the kinetic energy terms in Eq. 1.1 can be expressed in terms of $\mathbf{p}_\rho$ and $\mathbf{p}_R$,

$$\frac{P_A{}^2}{m_A} + \frac{P_B{}^2}{m_B} = \frac{1}{\mu}\,p_\rho{}^2 + \frac{1}{M}\,p_R{}^2,$$

and the Schrödinger equation can be written as

$$\left[\frac{1}{2\mu}\,p_\rho{}^2 + \frac{1}{2M}\,p_R{}^2 + V(\rho) - E\right]\Psi(\boldsymbol{\rho}, \mathbf{R}) = 0.$$

If we invoke the quantum mechanical axiom that to the momentum observable $\mathbf{p}$ there corresponds the operator $-i\hbar\boldsymbol{\nabla}$, then the Schrödinger equation becomes

$$\left[\frac{-\hbar^2}{2\mu}\,\nabla_\rho{}^2 - \frac{\hbar^2}{2M}\,\nabla_R^2 + V(\rho) - E\right]\Psi(\boldsymbol{\rho}, \mathbf{R}) = 0. \tag{1.5}$$

Equation 1.5 is a partial differential equation in six variables, $\boldsymbol{\rho}$ and $\mathbf{R}$. To solve this equation, it is usual to assume the separability of the coordinates $\boldsymbol{\rho}$ and $\mathbf{R}$,

$$\Psi(\boldsymbol{\rho}, \mathbf{R}) = \Phi(\mathbf{R})\psi(\boldsymbol{\rho}) \qquad E = E_\rho + E_R.$$

When these separations are substituted into Eq. 1.5 we obtain

$$\Phi(\mathbf{R})\left[\frac{-\hbar^2}{2\mu}\,\nabla_\rho{}^2 + V(\rho) - E_\rho\right]\psi(\boldsymbol{\rho}) = -\psi(\boldsymbol{\rho})\left[\frac{-\hbar^2}{2M}\,\nabla_R{}^2 - E_R\right]\Phi(\mathbf{R})$$

which can only be true for all $\boldsymbol{\rho}$ and $\mathbf{R}$ provided

$$\left[\frac{-\hbar^2}{2\mu}\,\nabla_\rho{}^2 + V(\rho) - E_\rho\right]\psi(\boldsymbol{\rho}) = 0 \tag{1.6}$$

and

$$\left[\frac{-\hbar^2}{2M}\,\nabla_R{}^2 - E_R\right]\Phi(R) = 0. \tag{1.7}$$

Our mathematical problem has now reduced to solving a pair of uncoupled three-dimensional partial differential equations.

Problem 1. Given the kinetic energy of a three-particle system as

$$\left(\frac{p_1^{\,2}}{2m_1} + \frac{p_2^{\,2}}{2m_2} + \frac{p_3^{\,2}}{2m_3}\right)$$

in the laboratory frame, convert this expression into relative momenta.

Equation 1.6 describes the motion of one of the particles relative to the other and has precisely the same form as the equation of motion of a particle of mass μ in an external potential field V. Equation 1.7 tells us that the center-of-mass of the entire system moves like a free particle of mass M. Therefore, we have shown that the nonrelativistic motion of two particles can be broken up into two one-particle problems. The motion of the center-of-mass cannot be ignored in calculating the outcome of a collision experiment. In an experiment it is usual to bombard a target particle, initially at rest, with a projectile which carries the total energy of the system, $E = E_\rho + E_R$. That is, the energy of relative motion of the two particles, E_ρ, is different from the bombarding energy E of the experiment and the observed scattering actually depends on whether the struck particle or the center-of-mass is initially at rest.

The laboratory coordinate system is *defined* to be that frame of reference in which the target particle is initially at rest. The center-of-mass coordinate system is *defined* to be that frame in which the center-of-mass is *always* at rest. It is easier to perform calculations in the center-of-mass frame, because there are three less variables; consequently we work with Eq. 1.6.

1.1.2. Boundary Conditions [1]

The collision problem is reduced to the solution of the partial differential equations (1.6 and 1.7) subject to boundary conditions—it is these conditions that determine which physical problem is under consideration. Figure 1 is a schematic drawing of a scattering experiment in which the projectiles from a source are collimated and directed toward target particles, which are essentially at rest in the laboratory frame. After scattering, some of the reaction products are registered in detectors which count the number of particles of a given kind which arrive there. The principal role of the collimator is to shield the detector from incident particles coming directly from the source without having been scattered there by the target particles. The more perfect the collimation, the better approximation it is to assume that all the incident projectiles travel parallel to the Z-axis, indeed they define a Z-axis. In most scattering experiments great care is taken to ensure that all the

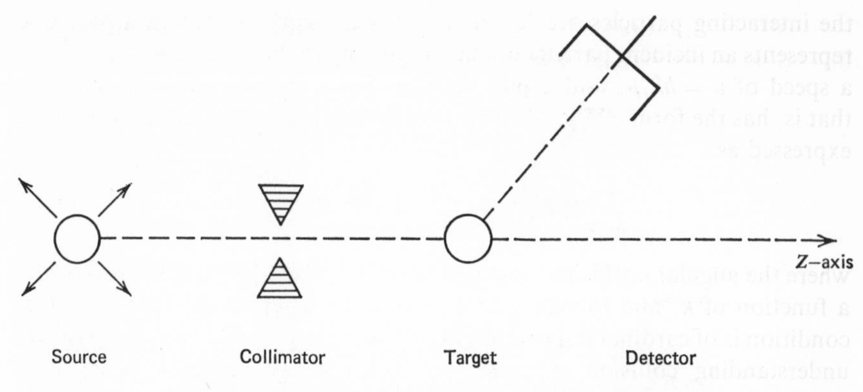

Figure 1 Schematic drawing of a scattering experiment.

projectiles have the same initial energy when they leave the source. In the following, it is assumed that the projectiles are sufficiently monochromatic and sufficiently collimated that we can assign a single wave number K to the incident beam and that the slight angular divergence from the collimator can be ignored.

If the target is absent, then the motion of the free projectile is described by Eq. 1.7 and no projectiles would be found outside the pencil emerging from the collimator; defining the wave number K by the relation $K^2 \equiv (2M/\hbar^2)E_R$, $K = Mv/\hbar$, then the solution of Eq. 1.7 must be

$$\Phi(\mathbf{R}) = e^{iKZ} \tag{1.8}$$

which represents a plane wave moving parallel to the Z-axis. Since $M(E_R)$ has dimensions of mass (energy), and $\hbar$ has dimensions of energy times time, then the units of K are $(\text{length})^{-1}$. We discuss the normalization of the plane wave later. Here, the center-of-mass is at the projectile itself. When the target is present, but there is no interaction between the target and the projectile, that is, $V(\rho)$ is absent, Eq. 1.7 still has the plane wave solution, furthermore the two particles are moving relative to each other parallel to the Z-axis and so the solution of Eq. 1.6 must be

$$\psi(\boldsymbol{\rho}) = e^{ikZ}, \qquad k^2 = \frac{2\mu}{\hbar^2} E_\rho, \qquad z = Z. \tag{1.9}$$

In other words, from the entire solution space of $(\nabla_\rho{}^2 + k^2)\psi = 0$ we have chosen those solutions which satisfy the boundary condition of moving parallel to the Z-axis.

In the presence of an interaction between the two particles, $V(\rho) \neq 0$, Eq. 1.8 remains unchanged, but now some particles will be registered at the detector due to the interaction between projectiles and target. That is, when the interacting particles are far apart we want $\psi(\boldsymbol{\rho})$ to contain a part that represents an incident particle of mass μ moving in the Z-direction, e^{ikZ}, with a speed of $v = \hbar k/\mu$, and a part representing a radially outgoing particle, that is, has the form $e^{ik\rho}/\rho$. Mathematically this asymptotic condition can be expressed as

$$\operatorname*{Lt}_{\rho\to\infty} \psi(\boldsymbol{\rho}) \sim N\left[e^{ikZ} + f(\theta, \phi)\frac{e^{ik\rho}}{\rho}\right], \tag{1.10}$$

where the angular coefficient function of $e^{ik\rho}/\rho$, namely $f(\theta, \phi)$, which is also a function of k, and therefore of E, has to be determined. This boundary condition is of cardinal importance in collision theory and forms the basis for understanding collision processes in which many different varieties of reaction products are generated in a collision process. In other words, of all possible solutions to the partial differential equation (1.6), we are only interested in those solutions which have the asymptotic form specified by Eq. 1.10.

1.1.3. Cross Sections

The first term in Eq. 1.10 represents an incident plane wave moving in the positive z-direction when we include the time factor, namely,

$$\Psi_{\text{inc}} \sim Ne^{ik(z-vt)}, \qquad |\Psi_{\text{inc}}|^2 = N^2.$$

From this normalization and the axiom of quantum mechanics that the wave function can be interpreted as a probability amplitude, we see that there are N^2 particles per unit volume. Hence, the number of particles crossing unit area in unit time is N^2v: this is called the incident flux of particles. Any experiment must measure this flux. Theoretically, it is possible to normalize the ingoing plane wave to have unit amplitude, that is, $N^2 = 1$, which gives an incident flux of v particles crossing unit area in unit time. Frequently the amplitude of the incident plane wave is normalized to ensure unit incident flux; in that case $N = v^{-1/2}$.

If we take $N^2 = 1$, then the number of particles per unit volume in the scattered wave at a very large distance ρ from the origin is

$$|\Psi_{\text{scatt}}|^2 = |e^{ik\rho}f(\theta, \phi)\rho^{-1}|^2 = \rho^{-2}\,|f(\theta\phi)|^2.$$

Consequently, the number of particles crossing an element of area, $(\rho \sin\theta \, d\phi) \times (\rho \, d\theta)$, per unit time is

$$v\rho^{-2}\,|f(\theta\phi)|^2\,\rho^2 \sin\theta \, d\theta \, d\phi = v\,|f(\theta\phi)|^2\,d\Omega,$$

where $d\Omega$ is an element of solid angle. We have now arrived at the most important concept of collision theory: the probability that a particle in the incident beam will be scattered into the solid angle $d\Omega$ is given by the ratio of

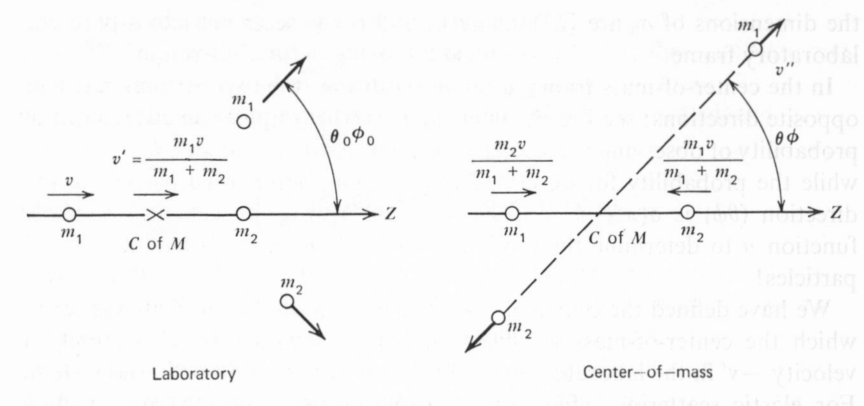

Figure 2 Schematic diagrams of binary collisions in the laboratory and center-of-mass frames.

the scattered to incident fluxes

$$v\,|f(\theta\phi)|^2\,d\Omega v^{-1} = |f(\theta\phi)|^2\,d\Omega \equiv d\sigma(\theta\phi). \tag{1.11}$$

The quantities, $f(\theta\phi)$, $|f(\theta\phi)|^2$, and $d\sigma(\theta\phi)$ are called the scattering amplitude, the scattering intensity, and the differential scattering cross section, and all depend on k, and therefore on E, although we have suppressed this variable in writing the functions. The total probability that a particle of energy E will be scattered in any direction by the interaction $V(\rho)$ is obtained by integrating over all angles

$$\int d\sigma(\theta\phi, E) = \int |f(\theta\phi, E)|^2\,d\Omega \equiv \sigma(E) \tag{1.12}$$

where the dependence of the scattering amplitude on the energy is now noted explicitly. The quantity $\sigma(E)$ is the total cross section.

From this brief discussion we see that the determination of $f(\theta\phi, E)$ leads to mathematical quantities that can be compared with experiment. Therefore, the theoretical problem is to solve Eq. 1.6, subject to the asymptotic boundary condition (1.10) and the usual requirement of solutions that are physically significant: we do not want solutions that are anywhere singular.

We now relate the cross section calculated in the center-of-mass frame to the cross section as calculated in the laboratory frame [2]. In the laboratory frame, suppose we have n scattering centers and an incident flux of N particles per sec per unit area, and suppose we count the number of particles which emerge into a solid angle $\Delta\omega_0$ around polar angles $\theta_0\phi_0$ in unit time. This number will be proportional to $(nN\,\Delta\omega_0)$, where the constant of proportionality depends on $\theta_0\phi_0$ and is written $\sigma_0(\theta_0\phi_0)$. Since the number per unit time has dimension $[T^{-1}]$ and $(nN\,\Delta\omega_0)$ has dimensions $[L^{-2}T^{-1}]$, then the dimensions of σ_0 are $[L^2]$, an area, and is the scattered intensity in the laboratory frame.

In the center-of-mass frame, after the collision, the two particles move in opposite directions; see Fig. 2. As in the laboratory frame, we can define the probability of observing the scattered incident particle at angle $\theta\phi$: it is $\sigma(\theta\phi)$, while the probability for observing the recoiling bombarded particle in the direction $(\theta\phi)$ is $\sigma(\pi - \theta, \phi + \pi)$. In other words, we can use the same function σ to determine the cross section for both the scattered and recoil particles!

We have defined the center-of-mass frame to be that coordinate system in which the center-of-mass is always at rest. Consequently, if we apply a velocity $-\mathbf{v}'$ in the laboratory system we arrive at the center-of-mass system. For elastic scattering, after the collision, m_1 continues to move with a velocity $m_2 v/(m_1 + m_2) \equiv v''$ in the center-of-mass frame. Relative to an observer at rest in the laboratory frame, m_1 moves with a velocity which is the vector resultant, $\mathbf{v}_1$, of m_1 relative to center-of-mass and of the velocity of the center-of-mass relative to laboratory; see Fig. 3. This resultant has already been defined to be at $(\theta_0\phi_0)$ relative to the direction of the incident beam. If there is no component of momentum perpendicular to the paper initially, then the scattering plane is that of the paper. Hence

$$\begin{aligned} \phi &= \phi_0 \\ v_1 \cos\theta_0 &= v' + v'' \cos\theta \\ v_1 \sin\theta_0 &= v'' \sin\theta \end{aligned} \tag{1.13}$$

When v_1 is eliminated from these equations, we obtain

$$\tan\theta_0 = \frac{\sin\theta}{\gamma + \cos\theta}; \qquad \gamma \equiv \frac{v'}{v''} = \frac{m_1}{m_2}. \tag{1.14}$$

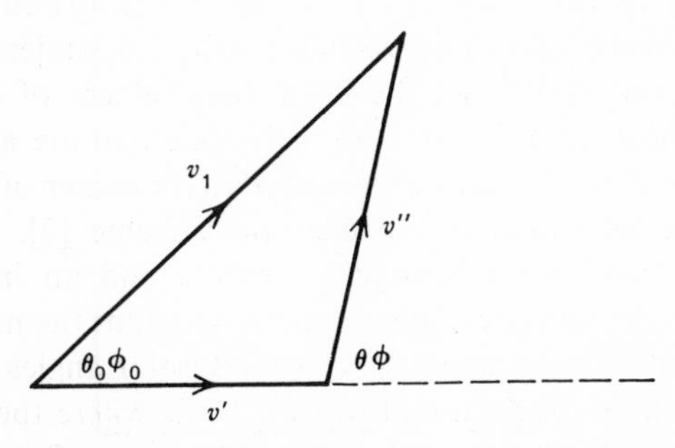

Figure 3 Vector addition diagram showing the relation between laboratory and center-of-mass scattering angles.

Problem 2. Generalize these results to the problem where particles of masses m_3 and m_4 emerge. If Q denotes the amount of energy which is converted from internal energy to kinetic energy of the emergent particles, show

$$\gamma = \left(\frac{m_1 m_3}{m_2 m_4} \times \frac{E}{E+Q}\right)^{1/2}, \qquad E = \frac{m_1 m_2 v^2}{2(m_1 + m_2)}.$$

From the definitions of the differential cross sections in the two frames we must have the same number of particles scattered into an element of solid angle, that is,

$$\sigma_0(\theta_0\phi_0)\, d\omega_0 = \sigma(\theta\phi)\, d\omega. \tag{1.15}$$

From the relationship between θ and θ_0 given in Eq. 1.14 we can construct the triangle drawn in Fig. 4.

From Fig. 4 we see immediately that

$$\cos\theta_0 = (\gamma + \cos\theta)(1 + 2\gamma\cos\theta + \gamma^2)^{-1/2}.$$

When this expression is differentiated with respect to θ_0, we obtain

$$-\sin\theta_0\, d\theta_0 = \frac{\sin\theta\, d\theta}{(1 + 2\gamma\cos\theta + \gamma^2)^{3/2}} \times [\gamma^2 + \gamma\cos\theta - \gamma^2 - 2\gamma\cos\theta - 1]. \tag{1.16}$$

This result is substituted into Eq. 1.15 written in the form

$$\sigma_0(\theta_0\phi_0)\sin\theta_0\, d\theta_0\, d\phi = \sigma(\theta\phi)\sin\theta\, d\theta\, d\phi \tag{1.17}$$

to give the relation between the scattered intensities in the two frames of

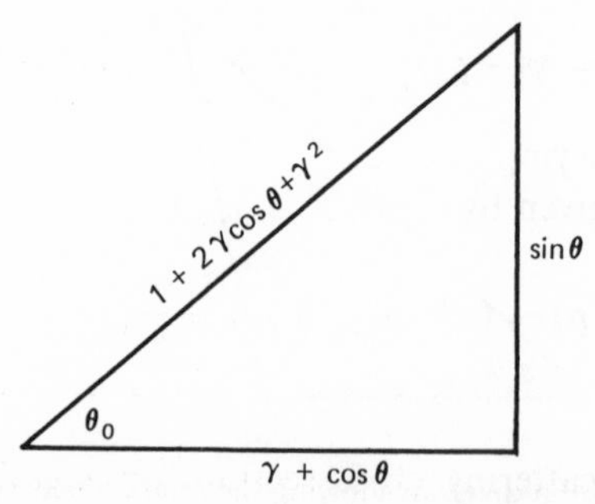

Figure 4 Thumbnail sketch of relations between $\sin\theta_0$, $\cos\theta_0$, and $\tan\theta_0$.

reference to be

$$\sigma_0(\theta_0\phi_0) = \frac{\sigma(\theta\phi)[1 + 2\gamma\cos\theta + \gamma^2]^{3/2}}{|1 + \gamma\cos\theta|}. \tag{1.18}$$

Since the total number of collisions must be independent of the manner we use to compute it, then the total cross section is the same for both laboratory and center-of-mass systems, but the energy must be transferred appropriately.

For equal mass particles, $\gamma = 1$, we have

$$\tan\theta_0 = \frac{\sin 2\cdot\theta/2}{1 + \cos 2\cdot\theta/2} = \frac{2\sin\theta/2\cdot\cos\theta/2}{2\cos^2\theta/2}$$
$$= \tan\theta/2. \tag{1.19}$$

Since we have proved the important result that for equal mass particles, not necessarily identical particles, $\theta_0 = \theta/2$, we see that as the polar angle in the center-of-mass system, θ, spans the range from 0 to π, θ_0 merely spans the range 0 to $\pi/2$; therefore no particles are *observed* in the backward hemisphere and Eq. 1.18 becomes

$$\sigma_0(\theta_0\phi_0) = \frac{\sigma(\theta\phi)(1 + 2\cos\theta + 1)^{3/2}}{|1 + \cos\theta|}$$
$$= \sigma(2\theta_0, \phi_0)2^{3/2}(1 + \cos\theta)^{1/2}$$
$$= \sigma(2\theta_0, \phi_0)4\cos\theta_0. \tag{1.20}$$

We now consider the cross section formulas for the collision between two *identical* particles, neglecting the effects of spin. The interchange of two identical particles does not affect **R**, but it does change the sign of the relative position vector $\boldsymbol{\rho}$. What we want to know is the effect of symmetry or antisymmetry of the spatial part of the wave function on the elastic scattering of a particle by a particle which is identical with it. Equation 1.10 is the asymptotic form for the unsymmetrized scattering wave function in the center-of-mass frame. The effect of the reflection operation, Π is

$$\Pi\psi(\boldsymbol{\rho}) = \psi(-\boldsymbol{\rho}) \underset{|\boldsymbol{\rho}|\to\infty}{\sim} e^{-ikz} + e^{ik\rho}/\rho f(\pi - \theta, \phi + \pi). \tag{1.21}$$

Consequently, the asymptotic forms of the symmetric and antisymmetric wave functions are given by

$$\Psi^{\pm} = \psi(\boldsymbol{\rho}) \pm \psi(-\boldsymbol{\rho}) \sim (e^{ikz} \pm e^{-ikz}) + \frac{e^{ik\rho}}{\rho}[f(\theta\phi) \pm f(\pi - \theta, \phi + \pi)] \tag{1.22}$$

The differential scattering cross section for a collision in which *either* particle is deflected into the solid angle $d\omega$ in the center-of-mass frame is the square of the magnitude of the coefficient of $e^{ik\rho}/\rho$; that is,

$$\sigma(\theta\phi)\,d\omega = |f(\theta\phi) \pm f(\pi - \theta, \phi + \pi)|^2\,d\omega. \tag{1.23}$$

The probability that a scattered particle will be found in the solid angle $d\omega_1$ and the particle with which it collided in the element of solid angle $d\omega_2$ is

$$\sigma(\theta\phi)\,d\omega_1\,d\omega_2.$$

If a collision is observed, it is impossible to tell after the collision which is the incident particle and which the target particle. When the interference term is neglected, $\sigma(\theta\phi)$ becomes the sum of the scattered intensities for observation of the incident particle, $|f(\theta\phi)|^2$, and of the struck particle, $|f(\pi - \theta, \phi + \pi|^2$.

The incident wave for the symmetric wave function, Ψ^+, may be written as (see the terms in parentheses in Eq. 1.22)

$$2 \cos kz = 2 \cos k(z_1 - z_2), \tag{1.24}$$

where z_i are the z-coordinates of particles 1 and 2, with an average value of $|\Psi^+_{\text{inc}}|^2$ of two and so the wave represents one particle crossing unit area in each beam (assuming particles 1 and 2 are fired at each other).

The differential cross section for a collision in which a particle is scattered through an angle θ_0 into a solid angle $d\omega_0$ in the laboratory coordinate system is, from Eq. 1.20,

$$\begin{aligned}\sigma_0(\theta_0\phi_0)\,d\omega_0 &= d\sigma_0(\theta_0\phi_0) = \sigma(\theta\phi)4\cos\theta_0\,d\omega_0\\ &= |f(\theta) \pm f(\pi - \theta)|^2\,4\cos\theta_0\,d\omega_0\\ &= |f(2\theta_0) \pm f(\pi - 2\theta_0)|^2\,4\cos\theta_0\,d\omega_0.\end{aligned} \tag{1.25}$$

This is the probability that a particle will be *observed* moving in a direction making an angle θ_0 with the direction of motion of the incident beam.

1.1.4. Partial Waves

Let us return now to how the three-dimensional partial differential equation given in Eq. 1.6 can be solved. If we make a mental picture of the scattering process in ordinary space, then it is self-evident that spherical polar coordinates are the natural coordinates in which to express the Laplacian, hence

$$\left\{\frac{-\hbar^2}{2\mu}\left[\rho^{-2}\frac{\partial}{\partial\rho}\rho^2\frac{\partial}{\partial\rho} + (\rho^2\sin\theta)^{-1}\frac{\partial}{\partial\theta}\sin\theta\frac{\partial}{\partial\theta} + (\rho^2\sin^2\theta)^{-1}\frac{\partial^2}{\partial\phi^2}\right] + V(\rho) - E_\rho\right\}\psi(\boldsymbol{\rho}) = 0.$$

Using the method of separation of variables, we can replace this three-dimensional partial differential equation by three sets of ordinary, decoupled, second-order differential equations

$$\left[\frac{d^2}{d\phi^2} + m^2\right]\Phi(\phi) = 0$$

$$\left[(1 - x^2)\frac{d^2}{dx^2} - 2x\frac{d}{dx} + l(l+1) - \frac{m^2}{1 - x^2}\right]\Theta(\theta) = 0,\ x = \cos\theta. \quad (1.26)$$

$$\left[\frac{1}{\rho^2}\frac{d}{d\rho}\rho^2\frac{d}{d\rho} + k^2 - \frac{l(l+1)}{\rho^2} - \frac{2\mu}{\hbar^2}V(\rho)\right]R_l(\rho) = 0.$$

Each set is infinite in number, since $|m| \leq l$ and $l = 0, 1, 2, 3, \ldots$. The Θ-equation is the Legendre equation and the angular functions are defined to be

$$Y_{lm}(\theta\phi) \equiv \Theta_l(\theta)\Phi_m(\phi) \quad (1.27)$$

with the following properties, which we use repeatedly in later sections, namely,

$$Y_{lm}(\theta\phi)^* = (-1)^m Y_{l-m}(\theta\phi), \qquad Y_{l0}(\theta\phi) = \left[\frac{2l+1}{4\pi}\right]^{1/2} P_l(\cos\theta),$$

$$\int Y_{lm}(\theta\phi)^* Y_{l'm'}(\theta\phi)\, d\Omega = \delta_{ll'}\,\delta_{mm'}. \quad (1.28)$$

We see later that the surface harmonics, $Y_{lm}(\theta\phi)$, are the eigenfunctions of a rigid rotator, which is of considerable interest in molecular problems. One of the most important properties of the Legendre polynomials is the addition theorem which relates the Legendre polynomials of the angle, Θ, between two vectors $\mathbf{r}$ and $\mathbf{r}'$, to the surface harmonics of the polar angles $\theta\phi$ and $\theta'\phi'$ of these two vectors, that is,

$$P_l(\cos\Theta) = \frac{4\pi}{(2l+1)}\sum_{m=-l}^{l} Y_{lm}^*(\theta\phi)Y_{lm}(\theta'\phi'). \quad (1.29)$$

Since the radial equation is independent of m, then the problem has axial symmetry, that is, no ϕ-dependence. Consequently, the wave function can be considered to be expanded in terms of the Legendre polynomials, $P_l(\cos\theta)$, as basis functions:

$$\psi(\rho\theta) = \sum_{l=0}^{\infty}\alpha_l P_l(\cos\theta)R_l(\rho), \quad (1.30)$$

where the constants α_l are to be chosen such that $\psi(\rho, \theta)$ has the asymptotic form (1.10). This latter statement underscores the fact that the solutions we generate must satisfy boundary conditions to be prescribed prior to our starting the solution process. By using the expansion in Eq. 1.30, the wave function is decomposed into a countably infinite set of radial functions. The functions $R_l(\rho)$ are called the partial waves, with the spectroscopic notation

$$l = 0, \quad s\text{-waves}; \qquad l = 1, \quad p\text{-waves}; \quad l = 2, \quad d\text{-waves, and so on.}$$

We take note of the fact that the "cost" of reducing the solution of (1.6), a three-dimensional partial differential equation, to the one-dimensional equation (1.26) for $R_l(\rho)$, is the need, in principle, to solve an infinite number of R-equations. In practice, we hope that the expansion (1.30) converges quickly. Only with the advent of the digital computer has it been possible to examine such convergence questions.

Equation 1.26 is a second-order ordinary differential equation and as such has two linearly independent solutions. The general solution of this equation is obtained by taking a linear superposition of these independent solutions, the coefficients in this combination being determined by the boundary conditions. It is now appropriate to point out a fundamental difference between the scattering problem under discussion here and the bound-state problem. In the former, k^2 is known In the latter, k^2, the eigenvalue, is unknown and is to be determined by the solution process. In both problems, the radial equation is homogeneous in R; consequently, whatever solution we extract we can multiply it by a ρ-independent constant and still have a solution. In other words, the normalization of R is arbitrary, which implies that one of the two integration constants can be assigned arbitrarily. The remaining integration constant must be adjusted to ensure that the boundary conditions are fulfilled—the explicit procedure is given in Section 1.4.

Let us recast the radial equation in a slightly more tractable form by writing

$$R_l(\rho) = \frac{F_l(k\rho)}{k\rho,} \tag{1.31}$$

which leads to F satisfying

$$\left[\frac{d^2}{d\rho^2} + k^2 - \frac{l(l+1)}{\rho^2} - \frac{2\mu V(\rho)}{\hbar^2}\right] F_l(k\rho) = 0. \tag{1.32}$$

This equation is the basic mathematical equation for the whole discussion presented in Chapter 1. These equations for F_l, $l = 0, 1, 2, 3, \ldots$, are said to be uncoupled from one another since we can solve for F_{l_1} without considering the solution F_{l_2}, where $l_2 \neq l_1$. In Section 1.1.2 we showed that the dimensions of k^2 were inverse length squared, which, by inspection, we can see is consistent with the other three operators in Eq. 1.32. In numerical calculations it is convenient to work with dimensionless quantities, so Eq. 1.32 is multiplied throughout by $\alpha_0{}^2$, where α_0 is the Bohr radius. If we let $r \equiv \rho/\alpha_0$ be the independent variable and still use k^2 to represent the now dimensionless quantity, then Eq. 1.32 becomes

$$\left[\frac{d^2}{dr^2} + k^2 - \frac{l(l+1)}{r^2} - 2U(r)\right]F_l(kr) = 0, \tag{1.33}$$

where

$$U(r) \equiv \frac{\mu a_0{}^2 V(r)}{\hbar^2}.$$

It is instructive to describe the influence of the four terms which make up the operator in square brackets in Eq. 1.33. For zero-energy projectiles, $k^2 = 0$, and for regions of r where $U(r)$ is negligible, then Eq. 1.33 reduces to

$$\left[\frac{d^2}{dr^2} - \frac{l(l+1)}{r^2}\right]F_l(r) = 0.$$

Problem 3. Obtain the two linearly independent solutions to the preceding equations. Show that the s-wave solutions are straight lines.

The energy point $k^2 = 0$ is called the elastic scattering threshold. The behavior of the radial functions $F_l(kr)$ in the vicinity of thresholds is of considerable physical interest—this is described in detail in subsequent sections.

Away from the elastic scattering threshold, $k^2 > 0$, but in the region of configuration space where $U(r)$ can still be neglected, Eq. 1.33 becomes

$$\left[\frac{d^2}{dr^2} + k^2 - \frac{l(l+1)}{r^2}\right]F_l(kr) = 0, \tag{1.34}$$

the solutions being the well-known spherical Bessel functions

$$krj_l(kr) \quad \text{and} \quad krn_l(kr).$$

The centrifugal barrier, that is, the term $l(l+1)/r^2$, is a repulsive potential, and as l increases the slope of j_l at the origin becomes progressively smaller. As k^2 increases, the frequency of the oscillations in the Bessel functions increases also. In summary, when $U(r)$ can be neglected, the F-functions are well-known analytic solutions.

In those regions of r where $U(r)$ cannot be neglected then we must solve Eq. 1.33, and usually the solutions are not the well-known functions of classical analysis. For example, in Fig. 5 we present the potential as "seen" by electrons moving in the vicinity of atomic nitrogen in its ground state. It is evident from this figure that $U(r)$ is attractive and is dominant in the intermediate regions of configuration space, while the repulsive centrifugal potential is dominant at small and large values of r. We see that the interplay

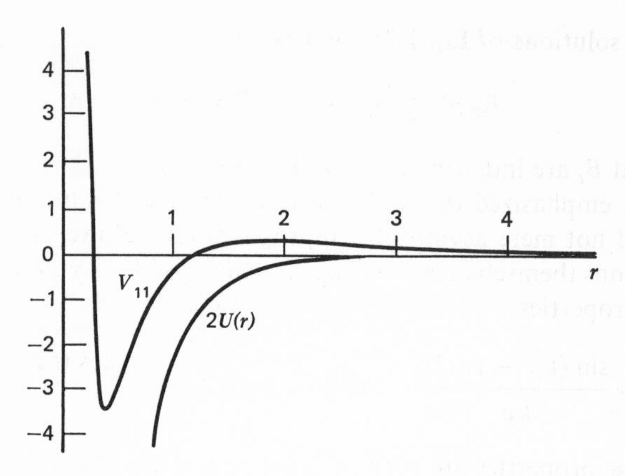

Figure 5 The full static potential is V_{11} (including the centrifugal barrier) for electrons incident on atomic nitrogen.

between these two potentials generates a barrier at about $1.5\alpha_0$, whose significance is discussed later.

In a scattering experiment, the collector is placed far from the interaction region; in other words, r is very large. The region of space where r is extremely large is called the asymptotic region and much of the discussion in scattering theory is concerned with the characteristics of the F-functions in the asymptotic region. When the interacting particles are infinitely far apart, then we should expect there would be no interaction between them, and

$$\lim_{r\to\infty} U(r) \to 0, \qquad \lim_{r\to\infty} F_l(kr) \to f_l(kr)$$

where f_l satisfies Eq. 1.34.

In actual fact, we can see that $F_l(kr)$ will not have the spherical Bessel functions of integer order as its asymptotic limits, if $U(r) \to 0$ like γ/r^2, where γ is some constant. It is clear that the third term in Eq. 1.34 should be

$$\frac{l(l+1)+\gamma}{r^2}.$$

Indeed, if $U(r) \to 0$ like $2\eta/r$, where η is some constant, then in the asymptotic domain $U(r)$ will have a more profound influence on the characteristics of F_l than the centrifugal barrier. For such potentials we must work with Coulomb functions rather than spherical Bessel functions, and this is discussed later in Section 1.1.5. For the remainder of this section we consider only potentials which vanish in the asymptotic region faster than r^{-2}. We have then, that at large values of ρ, R must be a linear superposition of the only two independent solutions of Eq. 1.34, namely,

$$R_l(\rho) \underset{\rho\to\infty}{\sim} A_l j_l(k\rho) - B_l n_l(k\rho) \tag{1.35}$$

where A_l and B_l are independent of ρ, but not of k!

It must be emphasized that krj_l and krn_l are indeed solutions for all r of Eq. 1.34 and not mere asymptotic approximations. However, the spherical Bessel functions themselves can be approximated in the asymptotic region by noting the properties

$$j_l(k\rho) \underset{\rho\to\infty}{\sim} \frac{\sin(k\rho - l\pi/2)}{k\rho} \quad \text{and} \quad n_l(k\rho) \underset{\rho\to\infty}{\sim} -\frac{\cos(k\rho - l\pi/2)}{k\rho}.$$

When these properties are used in Eq. 1.35 and the result substituted into the left-hand side of Eq. 1.31 we find the asymptotic result that

$$F_l(\rho) \underset{\rho\to\infty}{\sim} A_l \sin\left(k\rho - \frac{l\pi}{2}\right) + B_l \cos\left(k\rho - \frac{l\pi}{2}\right).$$

This result can be written in the alternative form

$$F_l(\rho) \sim C_l \sin\left[k\rho - \frac{l\pi}{2} + \delta_l(k)\right]; \tag{1.36}$$

where

$$\tan\delta_l(k) \equiv \frac{B_l(k)}{A_l(k)},$$

$C_l(k)$ is a normalization constant, and $\delta_l(k)$ is called the lth partial wave phase shift. The concept of a phase shift is a cornerstone of current non-relativistic scattering theory. The question as to what has suffered a shift in phase is now considered.

In Section 1.1.3 we defined the incident and scattered portions of the full wave function in the asymptotic domain,

$$\psi(\rho, \theta) \underset{\rho\to\infty}{\sim} \psi_{\text{inc}} + \psi_{\text{scatt}},$$

so

$$\psi_{\text{scatt}} = \psi(\rho, \theta) - e^{ik\rho\cos\theta}, \quad \text{since} \quad z = \rho\cos\theta$$

The plane wave can also be decomposed into a denumerably infinite set of partial waves [3]

$$e^{ik\rho\cos\theta} = \sum_{l=0}^{\infty} (2l+1) i^l j_l(k\rho) P_l(\cos\theta); \tag{1.37}$$

hence using Eq. 1.30 for the full-wave function, we have the following partial-wave expansion for the scattered wave

$$\psi_{\text{scatt}} = \sum_{l=0}^{\infty} \{\alpha_l R_l(\rho) - (2l+1)i^l j_l(k\rho)\} P_l(\cos\theta). \tag{1.38}$$

In order to determine the r-independent expansion coefficients, $\alpha_l(k)$, we can take advantage of our knowledge of the solutions at large r. In the asymptotic domain we substitute the sine and cosine forms of R_l and j_l and rewrite them in terms of radial ingoing, $\rho^{-1}e^{-ik\rho}$, and outgoing, $\rho^{-1}e^{ik\rho}$, waves, that is, convert the sine and cosine functions to exponentials. The central point of the argument is that since ψ_{scatt} can have outgoing waves only, then the coefficient of the ingoing wave vanishes, giving

$$\alpha_l C_l = (2l+1)i^l e^{i\delta_l}. \tag{1.39}$$

This result is substituted into Eq. 1.38 to give

$$\psi_{\text{scatt}} \underset{\rho\to\infty}{\sim} \frac{e^{ik\rho}}{\rho} \sum_{l=0}^{\infty} \frac{(2l+1)}{2ik} [e^{2i\delta_l} - 1] P_l(\cos\theta). \tag{1.38a}$$

However, in Eq. 1.10, the fundamental boundary condition, we have

$$\psi_{\text{scatt}} \underset{\rho\to\infty}{\sim} \frac{e^{ik\rho}}{\rho} f(\theta),$$

Comparison of these last two relations gives an explicit formula for the scattering amplitude in terms of the phase shift!

$$f(\theta) = \sum_{l=0}^{\infty} \frac{(2l+1)[e^{2i\delta_l} - 1] P_l(\cos\theta)}{2ik}. \tag{1.40}$$

In other words, if we can determine the partial wave phase shifts and, furthermore, if the series in l is fairly rapidly convergent, then we can compute $f(\theta)$ and therefore the cross sections, using the expressions in the previous subsection.

What is involved in determining $\delta_l(k)$? The incident monochromatic, well-collinated beam of projectiles has been represented by a plane wave moving in the positive z-direction. This plane wave has been decomposed into partial waves, and we have to compute the change in phase in each of those partial waves by the potential. In general, this involves the numerical solution of the second-order ordinary differential equation (1.33); various methods are described in detail in Section 1.4.

We conclude this section by discussing three functions which are commonly used in collision theory. Writing Eq. 1.30 in the asymptotic region as

$$\psi(\rho, \theta) \underset{\rho\to\infty}{\sim} -\sum_l \frac{(2l+1)i^l P_l(\cos\theta)}{2ik\rho} [e^{-i(k\rho - l\pi/2)} - S_l(k)e^{i(k\rho - l\pi/2)}] \tag{1.41}$$

where

$$S_l(k) \equiv e^{2i\delta_l(k)},$$

and is called the single-channel "S-matrix," that is, the coefficients of the outgoing spherical wave, when the ingoing spherical wave is normalized to unit amplitude. With this definition, we can rewrite the scattering amplitude formula given in Eq. 1.40 in the form

$$f(\theta) = \frac{1}{2ik}\sum_l (2l+1)(S_l - 1)P_l(\cos\theta)$$

$$\equiv \frac{1}{2ik}\sum_l (2l+1)T_l P_l(\cos\theta) \tag{1.42}$$

which defines the transition, or **T**-matrix, for a single channel. The total cross section is given by taking the modulus squared of Eq. 1.42 and integrating over all angles to obtain

$$\sigma(k^2) = \tfrac{1}{4}k^2 \sum_{l=0}^{\infty}\sum_{l'=0}^{\infty}(2l+1)(2l'+1)T_l^* T_{l'}$$

$$\times\, 2\pi \int_0^{\pi} P_l(\cos\theta)P_{l'}(\cos\theta)\sin\theta\, d\theta$$

$$= \frac{\pi}{k^2}\sum_{l=0}^{\infty}(2l+1)\,|T_l|^2 \tag{1.43}$$

We have used the orthonormality of the Legendre polynomials to eliminate one of the sums.

Finally, we can rewrite Eq. 1.30, using Eqs. 1.36 and 1.39, in the form

$$\psi(\rho,\theta) \sim \sum_l (2l+1)i^l e^{i\delta_l}\cos\delta_l P_l(\cos\theta)\frac{1}{k\rho}$$

$$\times\, [\sin(k\rho - \tfrac{1}{2}l\pi) + R_l(k)\cos(k\rho - \tfrac{1}{2}l\pi)]$$

where the **R**, or reactance, matrix is defined simply by

$$R_l \equiv \tan\delta_l. \tag{1.44}$$

The many-channel analog of this quantity plays a central role in our discussions in Chapter 2; its principal virtue, in contrast to S_l and T_l, is that it is a real quantity when $\delta_l(k)$ is real.

1.1.5. Coulomb Scattering†

The Coulomb potential

$$\frac{ZZ'e^2}{r}$$

† See Schiff [2], Section 20.

between two particles of electric charge Ze and $Z'e$ is of great importance in atomic collisions. We have shown how this potential is an exception to the techniques developed in the preceding section for calculating cross sections. With the Coulomb potential, Eq. 1.6 becomes

$$\left[\nabla_\rho^2 - \frac{2\alpha k}{\rho} + k^2\right]\psi(\boldsymbol{\rho}) = 0, \qquad \alpha \equiv \frac{\mu ZZ'e^2}{\hbar^2 k}. \tag{1.45}$$

From our intuitive description of the scattering process, presented in the early section on boundary conditions, we saw that $\psi(\boldsymbol{\rho})$ is required to have the asymptotic form

$$\psi(\boldsymbol{\rho}) \sim I + f(\theta, \phi)S, \tag{1.46}$$

where I represents an incident wave and S a scattered wave. The objective of this section is to derive the explicit forms of I, S, and $f(\theta, \phi)$, when Eq. 1.45 is solved for all ρ, as well as when it is only valid beyond some value of ρ. In accomplishing these objectives, we encounter some functions of classical mathematical analysis whose properties are examined in some detail in order to familiarize the student with analytical techniques which are used extensively in calculating atomic cross sections.

The method described earlier for solving the three-dimensional partial differential equation, involved an expansion of $\psi(\boldsymbol{\rho})$ in terms of Legendre polynomials, $P_l(\cos\theta)$, thereby introducing the concepts of partial waves and phase shifts. In this section we show that these concepts are still valid and useful for Coulomb scattering.

The usual method for solving partial differential equations involving the Laplacian is by separation of variables. The choice of which set of variables should be used is governed by the physical characteristics of the problem under consideration. In the scattering problem, intuitively we should expect the projectile to follow a parabolic orbit; consequently, we introduce the parabolic coordinates

$$\zeta = r - z, \qquad \eta = r + z, \qquad \phi = \tan^{-1}\frac{y}{x}.$$

Due to the Coulomb potential being spherically symmetric, a single scattering event will take place in a plane; in other words, the wave function describing the scattering will be independent of ϕ. That is to say, we have $\psi(\zeta, \eta)$, which, from Eq. 1.10, we should expect to contain components whose dominant factors are asymptotic to e^{ikz} and e^{ikr}/r, but not to e^{-ikr}/r. We should expect

$$\psi(\zeta, \eta) = e^{ikz}F(\zeta) \tag{1.47}$$

to satisfy these criteria, but not $e^{ikz}F(\eta)$! When Eq. 1.47 is substituted into

Eq. 1.45 we obtain the second-order ordinary differential equation

$$\zeta \frac{d^2F}{d\zeta^2} + (1 - ik\zeta)\frac{dF}{d\zeta} - \alpha kF = 0. \tag{1.48}$$

From the theory of such equations, it is well known that there are two linearly independent solutions and the general solution can be constructed by taking a linear superposition of these solutions, the pair of constant coefficients in such a superposition being determined by the boundary conditions.

In the first instance, we solve Eq. 1.45, and therefore Eq. 1.48, for all ζ. This leads at once to a boundary condition at the origin—F must be regular at the origin in order that ψ be physically significant. This condition will determine one of the two integration constants. The second constant can be chosen arbitrarily since (1.48) is homogeneous in F and therefore, once we have a solution, it can be multiplied by any ζ-independent factor.

Equation 1.48 is very well known in classical mathematical analysis and the properties of its solutions are well established. However, in view of its importance in atomic collision problems, we investigate various properties in detail. We begin by rewriting Eq. 1.48 in the form

$$z\frac{d^2F}{dz^2} + (b - z)\frac{dF}{dz} - aF = 0, \tag{1.48a}$$

where $z \equiv ik\zeta$, $b \equiv 1$, and $a \equiv -i\alpha$.

To determine the behavior of the solutions near the origin, we try an ascending power series solution

$$F(a, b, z) = \sum_{p=0}^{\infty} d_p z^{p+\sigma}$$

and find that $\sigma = 0$ or $(1 - b)$. For $\sigma = 0$, the solution which is regular at the origin, and which is called the confluent hypergeometric function, we have

$$d_p = \frac{\Gamma(a + p)\Gamma(b)d_0}{\Gamma(a)\Gamma(b + p)p!},$$

which leads to (setting $d_0 = 1$, which we are free to do since Eq. 1.48a is homogeneous in F)

$$F(a, b, z) = (b - 1)!\sum_{n=0}^{\infty} \frac{c_n z^n}{(b + n - 1)!}$$

$$= 1 + \frac{a}{b \cdot 1}z + \frac{a(a + 1)}{b(b + 1)2!}z^2 + \cdots,$$

where c_n is the coefficient of x^n in the expansion of $(1 - x)^{-a}$. We have thus found the solution to Eq. 1.48 that we were looking for, and now we must investigate its asymptotic properties.

Problem 4. Derive the preceding recurrence relation for d_p.

At this point we refer the reader to Mott and Massey† for the technique of expressing (1.49) as an integral along any closed path encircling the origin once in an anticlockwise direction and using the resulting expression to reexpress the confluent hypergeometric function as a linear superposition of an alternative pair of linearly independent solutions of (1.48a), namely,

$$F(a, b, z) = W_1(a, b, z) + W_2(a, b, z) \tag{1.50}$$

where W_1 and W_2 have the integral forms

$$\begin{aligned} W_1 &= \frac{\Gamma(b)}{2\pi i}(-z)^{-a}\int_{\gamma_1}\left(1 - \frac{t}{z}\right)^{-a} e^t t^{a-b}\,dt \\ W_2 &= \frac{\Gamma(b)}{2\pi i}(z)^{a-b}e^z\int_{\gamma_1}\left(1 + \frac{t}{z}\right)^{a-b} e^t t^{-a}\,dt \end{aligned} \tag{1.51}$$

Let us expand the expressions in parentheses in the integrands and invert the order of performing the sum and integral to give

$$W_1(a, b, z) = \Gamma(b)(-z)^{-a}\sum_{n=0}^{\infty}\frac{\Gamma(a + n)}{\Gamma(a)n!\,z^n}\frac{1}{2\pi i}\int e^t t^{a-b+n}\,dt$$

The integral can be reexpressed in terms of the gamma function using the theorem

$$\frac{1}{\Gamma(x)} = \frac{1}{2\pi i}\int e^t t^{-x}\,dt$$

Hence we have

$$\begin{aligned} W_1(a, b, z) &= \frac{\Gamma(b)}{\Gamma(a)}(-z)^{-a}\sum_{n=0}^{\infty}\frac{\Gamma(a + n)z^{-n}}{\Gamma(b - a - n)n!} \\ &= \frac{\Gamma(b)}{\Gamma(b - a)}(-z)^{-a}G(a, a - b + 1, -z), \end{aligned} \tag{1.52}$$

where

$$G(\alpha, \beta, z) \equiv 1 + \frac{\alpha\beta}{z1!} + \frac{\alpha(\alpha + 1)\beta(\beta + 1)}{z^2 2!} + \cdots$$

† See Ref. 3, p. 58.

It is left to the reader to show that in a similar manner we have the following asymptotic expansion for W_2:

$$W_2(a, b, z) \sim \frac{\Gamma(b)}{\Gamma(a)} e^z z^{a-b} G(1 - a, b - a, z). \tag{1.53}$$

We now have all the mathematical apparatus necessary to go back to Eq. 1.47 and derive its asymptotic form and compare it with Eq. 1.46, the desired form. Equations 1.52 and 1.53 are substituted into Eq. 1.50 and the result is substituted into (1.47) to give

$$\psi(\zeta, \eta) \sim \frac{e^{ikz}(-ik\zeta)^{i\alpha}}{\Gamma(1 + i\alpha)} \left\{1 + \frac{\alpha^2}{ik\zeta}\right\} + \frac{e^{ikr}(ik\zeta)^{-i\alpha-1}}{\Gamma(-i\alpha)}.$$

We now make use of the following simple relations

$$\begin{aligned} (k\zeta)^{i\alpha} &= e^{i\alpha \log k\zeta} \\ (-i)^{i\alpha} &= e^{\pi\alpha/2} \\ \Gamma(1 - z) &= -z\Gamma(-z) \end{aligned} \tag{1.54}$$

to write

$$\psi(\zeta, \eta) \sim \frac{e^{\pi\alpha/2}}{\Gamma(1 + i\alpha)} \times \left[\left\{1 + \frac{\alpha^2}{ik(r - z)}\right\} e^{i(kz + \alpha \log k[r-z])} + \frac{\alpha\Gamma(1 + i\alpha)}{\Gamma(1 - i\alpha)} \frac{e^{i(kr - \alpha \log k[r-z])}}{k(r - z)}\right]$$

We have thus seen the precise manner by which the logarithmic phase factors, which always occur in Coulomb scattering problems, appear in the exponentials. This last expression is now beginning to take on the structure, (1.46), that we are looking for. To determine $f(\theta)$, we write

$$\frac{e^{-i\alpha \log k(r-z)}}{k(r - r\cos\theta)} = \frac{e^{-i\alpha \log 2kr \sin^2(\theta/2)}}{2kr \sin^2(\theta/2)} = \frac{e^{-i\alpha \log \sin^2(\theta/2)}}{2kr \sin^2(\theta/2)} e^{-i\alpha \log 2\alpha r}$$

and so, normalizing the incident wave to unit amplitude, we have the final result

$$\psi(\zeta, \eta) \sim \left\{1 + \frac{\alpha^2}{ik(r - z)}\right\} e^{ikz + i\alpha \log k(r-z)} + \left\{\frac{\Gamma(1 + i\alpha)\alpha e^{-i\alpha \log \sin^2(\theta/2)}}{\Gamma(1 - i\alpha) 2k \sin^2(\theta/2)}\right\} \frac{e^{ikr - i\alpha \log 2kr}}{r}, \tag{1.55}$$

or

$$\psi(\zeta, \eta) = e^{-\pi\alpha/2}\Gamma(1 + i\alpha)e^{ikz}F(-i\alpha, 1, ik\zeta) \tag{1.56}$$

The asymptotic form (1.55) has the structure of Eq. 1.46, and so gives the explicit form of I, S, and

$$f(\theta) = \frac{\alpha}{2k \sin^2(\theta/2)} \exp i\left[-\alpha \log \sin^2\left(\frac{\theta}{2}\right) + 2i\eta_0 + i\pi\right] \qquad (1.57)$$

where

$$e^{2i\eta_0} \equiv \frac{\Gamma(1+i\alpha)}{\Gamma(1-i\alpha)}.$$

Later we show that η_0 is the s-wave Coulomb phase shift. (The definition of η_0 has used the properties that $\Gamma(1 \pm i\alpha)$ have equal amplitudes and the phases are equal in magnitude but opposite in sign.)

From Eq. 1.55 we see that the Coulomb potential, as evidenced by the appearance of α, has distorted both the incoming plane wave and the outgoing scattered wave. The scattered intensity is given by

$$|f(\theta)|^2 = \frac{\alpha^2}{4k^2 \sin^4(\theta/2)}$$

which is the same formula Rutherford derived from classical mechanics.

To solve the three-dimensional partial differential equation (1.45), in terms of spherical polar coordinates (still ϕ-independent), we write

$$\psi(\rho, \theta) = \sum_{l=0}^{\infty} C_l R_l(\rho) P_l(\cos\theta), \qquad (1.58)$$

with the radial equation (1.26) taking the form

$$\left[\frac{1}{x^2}\frac{d}{dx}x^2\frac{d}{dx} + 1 - \frac{2\alpha}{x} - \frac{l(l+1)}{x^2}\right] R_l(x) = 0, \qquad x \equiv k\rho, \qquad (1.59)$$

where C_l is a normalization constant chosen to ensure that the partial wave expansion (1.58) is identical to the solution expressed earlier in parabolic coordinates.

We try for a solution by extracting from R_l some of its anticipated characteristics, introducing the substitution

$$R_l(x) = x^l e^{ix} F_l(x), \qquad (1.60)$$

and making a change of independent variable to arrive at

$$z\frac{d^2F}{dz^2} + (2l + 2 - z)\frac{dF}{dz} - (l + 1 + i\alpha)F = 0 \qquad (1.61)$$

which is of the form of (1.48a) with $b = 2l + 2$ and $a = (l + 1 + i\alpha)$. In other words, $F_l(x)$ is the confluent hypergeometric function, and we can derive the asymptotic form of the partial wave R_l using the properties of W_1

and W_2, namely,

$$R_l(kr) \sim \Gamma(2l+2)2^{-l-1}(k\rho)^{-1}(i)^{-i\alpha}\left[\frac{e^{ik\rho}i^{-l-1}(2k\rho)^{-i\alpha}}{\Gamma(l+1-i\alpha)} + \frac{e^{-ik\rho}i^{l+1}(2k\rho)^{i\alpha}}{\Gamma(l+1+i\alpha)}\right]$$

When the relations (1.54) are used on the right-hand side and the Coulomb l-partial wave phase shift is defined by

$$\frac{\Gamma(l+1+i\alpha)}{\Gamma(l+1-i\alpha)} \equiv e^{2i\eta_l}, \qquad \eta_l \equiv \arg\Gamma(l+1+i\alpha), \tag{1.62}$$

we obtain

$$R_l(k\rho) \sim \frac{\Gamma(2l+2)2^{-l}e^{\alpha\pi/2+i\eta_l}}{\Gamma(l+1+i\alpha)} \frac{\sin(k\rho - l\pi/2 - \alpha\log 2k\rho + \eta_l)}{k\rho} \tag{1.63}$$

Problem 5. Use the properties of the complex Γ-function to prove that Eqs. 1.62 are consistent definitions.

It is recalled that the normalization constants C_l are to be chosen such that the partial wave expansion (1.58) is identical to our solution expressed in parabolic coordinates, see Eq. 1.56; hence

$$e^{-\pi\alpha/2}\Gamma(1+i\alpha)e^{ik\rho\cos\theta}F(-i\alpha, 1, ik\rho(1-\cos\theta)) = \sum_{l'=0}^{\infty} C_{l'}R_{l'}(\rho)P_{l'}(\cos\theta)$$

When this identity is multiplied by $\sin\theta P_l(\cos\theta)$, the result integrated over all θ, and the orthogonality property of the Legendre functions invoked, we obtain

$$C_lR_l(\rho) = \frac{(2l+1)e^{-\pi\alpha/2}}{2}\Gamma(1+i\alpha) \times \int_0^\pi P_l(\cos\theta)e^{ik\rho\cos\theta}F(-i\alpha, 1, ik\rho(1-\cos\theta))\sin\theta\, d\theta.$$

When Eq. 1.37 is substituted for the exponential factor in the integrand

$$C_lR_l(\rho) = \frac{(2l+1)e^{-\pi\alpha/2}}{2}\Gamma(1+i\alpha)\sum_{n=0}^{\infty}(2n+1)i^n j_n(k\rho) \times \int_0^\pi P_l(\cos\theta)P_n(\cos\theta)F\sin\theta\, d\theta$$

and we use the first terms of the ascending power series expansion of j_n and F, namely

$$j_n(k\rho) \underset{\rho\to 0}{\sim} \frac{(2k\rho)^n n!}{(2n+1)!}, \qquad F \sim 1$$

we obtain the result, upon integration over θ,

$$C_l R_l(\rho) \underset{\rho \to 0}{\sim} e^{-\pi\alpha/2} \frac{\Gamma(1 + i\alpha)(2ik\rho)^l l!}{(2l)!} \tag{1.64}$$

Alternatively, from Eqs. 1.60 and 1.49 we have

$$R_l(x) \underset{x \to 0}{\sim} (k\rho)^l$$

which is substituted into Eq. 1.64 to yield

$$C_l \equiv e^{-\pi\alpha/2} \frac{\Gamma(1 + i\alpha) l! \, (2i)^l}{(2l)!}, \tag{1.65}$$

and Eq. 1.58 can be written as

$$\begin{aligned} \psi(\rho, \theta) &= \sum_{l=0}^{\infty} (2l + 1) i^l e^{i\eta_l} \left[e^{-i\eta_l - \pi\alpha/2} \frac{\Gamma(1 + i\alpha) l! \, 2^l R_l(\rho)}{\Gamma(2l + 2)} \right] P_l(\cos \theta) \\ &\equiv \sum_{l=0}^{\infty} (2l + 1) i^l e^{i\eta_l} L_l(\rho) P_l(\cos \theta) \end{aligned} \tag{1.66}$$

where the new radial function is chosen to be normalized as

$$\begin{aligned} L_l(\rho) &\equiv e^{-i\eta_l - \pi\alpha/2} \frac{\Gamma(1 + i\alpha) l! \, 2^l R_l(\rho)}{\Gamma(2l + 2)} \\ &\underset{\rho \to \infty}{\sim} \frac{\sin(k\rho - l\pi/2 - \alpha \log 2k\rho + \eta_l)}{k\rho}. \end{aligned} \tag{1.67}$$

In most atomic collision processes the scattering potential is not pure Coulombic. For example, at small radial distances, electron exchange effects are important. It is in the asymptotic region, where the long-range Coulomb potential is dominant. We can still expand the total wave function in terms of Legendre polynomials

$$\psi(\rho, \theta) = \sum_{l=0}^{\infty} G_l(\rho) P_l(\cos \theta) \tag{1.68}$$

where the new radial functions satisfy Eq. 1.59 for large ρ only. We recall that Eq. 1.59 has two linearly independent solutions, namely $L_l(\rho)$ which has been discussed above, and $K_l(\rho)$, the solution which is irregular at the origin, and which can be normalized such that

$$K_l(\rho) \underset{\rho \to \infty}{\sim} \frac{\cos(k\rho - l\pi/2 - \alpha \log 2k\rho + \eta_l)}{k\rho} \tag{1.69}$$

We know that asymptotically G_l must be a linear superposition of L_l and

K_l, namely,

$$G_l(\rho) \sim A_l L_l(\rho) + B_l K_l(\rho)$$
$$\sim C_l \sin(k\rho - l\pi/2 - \alpha \log 2k\rho + \eta_l + \sigma_l), \qquad \tan\sigma_l = \frac{B_l}{A_l}. \quad (1.70)$$

We see that the ratios of the coefficients of the irregular to regular Coulomb functions at the transition boundary, where the Coulomb potential alone is effective, measures a partial wave phase shift, σ_l due to the non-Coulomb part of the interaction potential. The normalization constant C_l must be chosen so that we still have the Coulomb modified incoming plane wave plus an outgoing spherical wave, namely,

$$\sum_l G_l(\rho) P_l(\cos\theta) \sim e^{ikz + i\alpha \log k(\rho - z)} + [f_c(\theta) + f_m(\theta)] \frac{e^{ik\rho - i\alpha \log 2k\rho}}{\rho} \quad (1.71)$$

where $f_c(\theta)$ is the Coulomb scattering amplitude, see Eq. 1.57, and we can expand the amplitude due to the non-Coulomb potential as

$$f_m(\theta) = \sum_l a_l P_l(\cos\theta) \quad (1.72)$$

Equations 1.70 and 1.66 are substituted into the left-hand and right-hand sides, respectively, of (1.71) to obtain

$$C_l \sin(k\rho - l\pi/2 - \alpha \log 2k\rho + \eta_l + \sigma_l)$$
$$= i^l(2l+1)e^{i\eta_l} \sin(k\rho - l\pi/2 - \alpha \log 2k\rho + \eta_l + \sigma_l) + a_l e^{i(k\rho - \alpha \log 2k\rho)}$$

which yields

$$C_l = i^l(2l+1)e^{i(\sigma_l + \eta_l)} \quad \text{and} \quad a_l = (2l+1)e^{2i\eta_l}(e^{2i\sigma_l} - 1), \quad (1.73)$$

and therefore $f_m(\theta)$.

1.1.6. Units

We have defined

$$k^2 = \frac{2\mu E}{\hbar^2},$$

and so

$$(ka_0)^2 = \frac{2\mu E a_0^{\,2}}{\hbar^2} = \frac{2\mu E}{\hbar^2}\left(\frac{\hbar^2}{\mu e^2}\right)^2 = \frac{E}{(\mu e^4/2\hbar^2)},$$

where a_0 is the first Bohr radius of the hydrogen atom. The energy of the ground state of the hydrogen atom is 1 rydberg and equals $\mu e^4/2\hbar^2$. Consequently, $(ka_0)^2$ is a number; the number of rydbergs of energy in E. If we multiply all our radial equations by $a_0^{\,2}$ we can then measure our radial distances in units of the Bohr radius, that is, $r = \rho/a_0$, and energies in rydbergs. Henceforth we understand k to mean ka_0. Phase shifts will be in radians.

In order to determine a set of units for the cross section we multiply numerator and denominator of the right-hand side of Eq. 1.43 by a_0^2 and we have

$$\frac{\sigma}{\pi a_0^2} = \frac{1}{k^2} \sum_{l=0}^{\infty} (2l + 1)\, |T_l|^2, \tag{1.74}$$

where k^2 is in rydbergs and the right-hand side is the cross section in units of πa_0^2.

A few remarks on the convention used here for the sign of the phase shift. We have seen that in the absence of the scattering potential, the solution which is regular at the origin is the spherical Bessel function. In other words, the incident plane wave is decomposed into ingoing and outgoing spherical waves with well-defined amplitudes. Qualitatively, the effect of the interaction can be thought of as distorting the shape of $j_l(kr)$—an oscillatory function. Crudely speaking, an attractive potential, $V(\rho) < 0$, will "pull in" the peaks of $j_l(kr)$, introducing what we define as a *positive* phase shift, $\delta_l > 0$. Conversely, a repulsive potential, $V(\rho) > 0$, will "push out" the peaks of $j_l(kr)$, introducing a *negative* phase shift; see Fig. 6. The phase shift is the distance along the abscissae between *corresponding* zeros of j_l and F_l, at sufficiently large r, that the distance between one pair of corresponding zeros and the next pair remains constant.

By and large, the experimental results presented here are drawn from two experimental areas: (1) photon absorption, in which it is usual to work with energy units in wavelengths (angstroms) or frequencies (cm^{-1}) and (2) particle scattering, in which the energies are given either in electron volts or

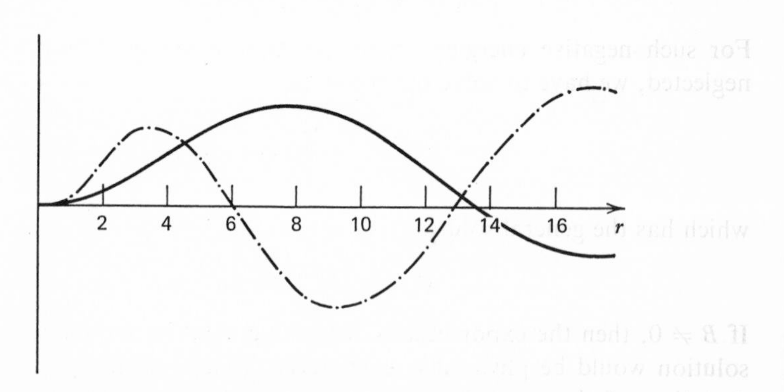

Figure 6 Comparison of j_2, continuous curve, with the radial function $F(^1D, k_1d)$ at $k_1^2 = 0.1783$ for $e - N^+1s^22s^22p^2(^1D)k_1d^2S^e$.

rydbergs (13.6 eV). In converting spectroscopic data from various experimental papers for comparison with the theoretical models we took 1 cm^{-1} = 1.24×10^{-4} eV. Resonance widths are always given in electron volts, and lifetimes, $\tau = \hbar/\Gamma$, in seconds.

1.2. RESONANCE PHENOMENA

1.2.1. Bound States

When a projectile is scattered from a target and the interaction can be represented by an attractive potential, then there is the possibility of the projectile being captured to form a bound compound system of projectile plus target. For s-wave scattering, that is, $l = 0$, the radial equation describing the relative motion of projectile and target (see Eq. 1.33), is

$$\left[\frac{d^2}{dr^2} + k^2 - 2U(r)\right]F(kr) = 0 \tag{1.75}$$

with the asymptotic boundary condition given in Eq. 1.41 to be

$$F(kr) \underset{r\to\infty}{\sim} C(k)[e^{-ikr} - S(k)e^{ikr}], \tag{1.76}$$

where $C(k)$ is some complex constant. Due to Eq. 1.75 being homogeneous in F, then we can set $C(k) = 1$. At the origin, we have $F(0) = 0$ a result we show in detail in Section 1.4.2.

In collision problems, k^2 is related to the energy of the projectile and for physically meaningful collision processes k^2 must be positive. We can enquire, however, as to the physical interpretation, if any, of solutions for negative k^2, that is, for

$$k = \pm i\kappa, \text{ with positive } \kappa.$$

For such negative energies, in the asymptotic region where $U(r)$ can be neglected, we have to solve the equation

$$\left[\frac{d^2}{dr^2} - \kappa^2\right]F(r) = 0,$$

which has the general solution

$$F(r) = Ae^{-\kappa r} + Be^{\kappa r}. \tag{1.77}$$

If $B \neq 0$, then the exponentially increasing solution dominates and such a solution would be physically meaningless because of the probability interpretation of the wave function. Consequently, any physically significant solution of the radial equation for $k^2 < 0$ must have $B \equiv 0$. Now consider

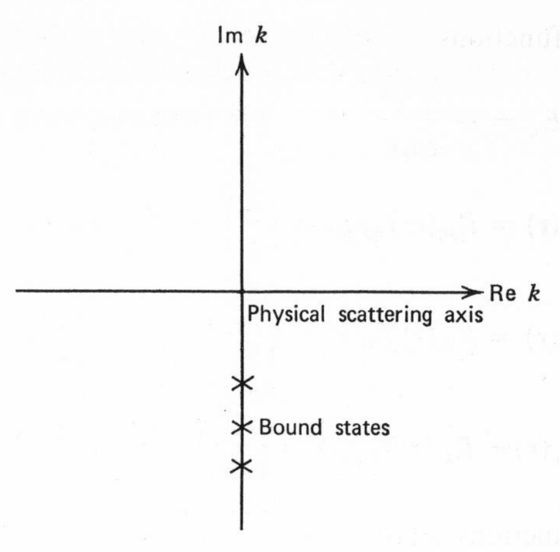

Figure 7 Complex k-plane for a single channel problem; crosses mark zeros of the S-matrix.

Eq. 1.76 for $k = -i\kappa$, and we have

$$F(r) \underset{r\to\infty}{\sim} e^{-\kappa r} - S(-i\kappa)e^{\kappa r}. \tag{1.78}$$

When this result is compared with Eq. 1.77 and we keep in mind our discussion on the physical significance of the solutions, we know that the differential equation can have a solution only at those negative k^2 for which the S-matrix has a zero

$$S(-i\kappa) = 0.$$

In other words, the location of the zeros of the S-matrix on the negative imaginary axis of the complex k-plane corresponds to solutions of (1.75) with

$$F(0) = 0 \qquad \text{and} \qquad F(r) \underset{r\to\infty}{\sim} e^{-\kappa r}$$

which are the familiar bound, or stationary, state solutions.

In summary: the analytical properties of the S-matrix correspond to the observable properties of quantum mechanical systems, in that zeros of S on the negative imaginary axis of the complex k-plane correspond to bound states, while the values of S on the real positive axis correspond to a continuous distribution of scattering states, see Fig. 7.

The most well-known attractive potential is the Coulomb potential, and the quantum mechanical treatment of the energy levels of the hydrogen atom supported by this potential is given in all wave mechanics texts.† The principal results of interest to use are the expressions for the energy levels and the

† See Schiff [2], Section (16).

first few wave functions

$$E_n = -\frac{Z^2e^2}{2a_0n^2} = -\frac{1}{n^2} \text{ in rydbergs, since } Z = 1 \tag{1.79}$$

$$\psi_{1s}(\mathbf{r}) = R_{10}(r)Y_{00}(\hat{\mathbf{r}}) = \left(\frac{Z}{a_0}\right)^{3/2} 2e^{-Zr/a_0}Y_{00}(\hat{\mathbf{r}})$$

$$\psi_{2s}(\mathbf{r}) = R_{20}(r)Y_{00}(\hat{\mathbf{r}}) = \left(\frac{Z}{2a_0}\right)^{3/2}\left(2 - \frac{Zr}{a_0}\right)e^{-Zr/2a_0}Y_{00}(\hat{\mathbf{r}})$$

$$\psi_{2p}(\mathbf{r}) = R_{21}(r)Y_{1m}(\hat{\mathbf{r}}) = \left(\frac{Z}{2a_0}\right)^{3/2}\frac{Zr}{a_0\sqrt{3}}e^{-Zr/2a_0}Y_{1m}(\hat{\mathbf{r}}).$$

These wave functions serve to remind us that bound state wave functions tend to zero exponentially as r tends to infinity. On the complex k-plane the bound states of the Coulomb potential will only be found at the denumerably infinite set of points $k = -in^{-1}$ crowding up on the origin as shown in Fig. 8.

A more familiar way of showing the bound states is by displaying the energy levels of the hydrogen atom as in Fig. 9.

Another potential of considerable physical interest is the dipole potential, $U(r) \approx -\beta r^{-2}$, $\beta > 0$, which has been studied in considerable detail [4].

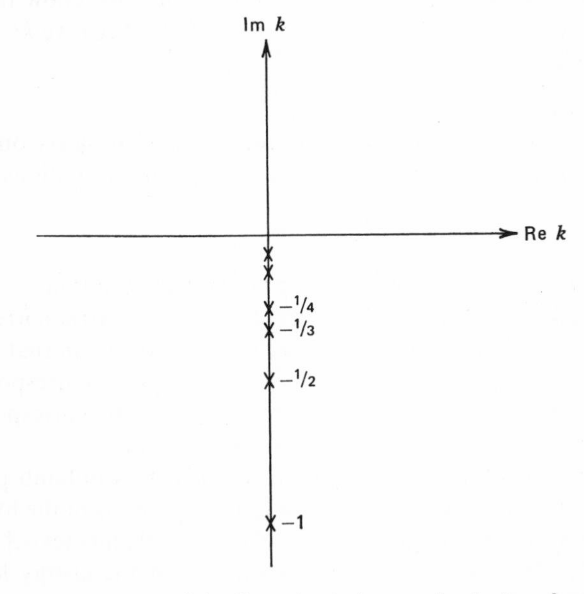

Figure 8 Location of the zeros of the S-matrix on the complex k-plane for the hydrogen atom.

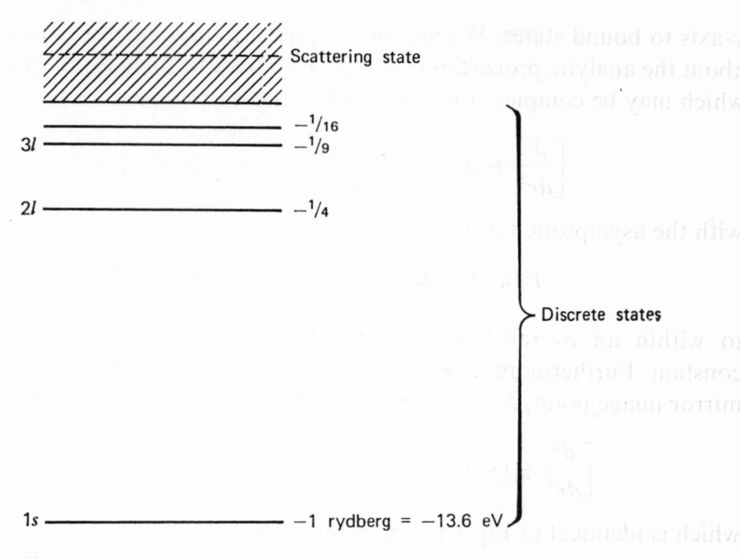

Figure 9 Schematic energy level diagram for the hydrogen atom.

This potential is of especial interest when we come to discuss the autoionizing levels of the two-electron systems H^- in Section 2.1.5, in which we show that the potential in the Schrödinger equation decreases as r^{-2} at large distances, but has a more complicated form at small distances. Landau and Lifshitz show that for certain values of l and β the discrete spectrum contains an *infinite* number of negative energy levels just like the Coulomb field.

Actual collision problems are not usually so simple. For example, in the scattering of electrons by He^+ we show later that the potential has the asymptotic form

$$U(r) = \frac{A}{r^2} - \frac{B}{r},$$

a mixture of Coulomb and dipole potentials. Landau and Lifshitz, p. 128, derive the formula for the infinite set of *equidistant* energy levels supported by this potential.

Problem 6. Given $\gamma = -l(l + 1) + \beta = -1 + \sqrt{37}$ as the coefficient of r^{-2} in the radial equation at large r and the potential has any form at small distances, discuss the form of the wave function and the negative energy levels.

We have been using the radial equation (1.33) together with its allied boundary condition (1.41) to relate zeros of $S_l(k)$ on the negative imaginary

k-axis to bound states. We use these equations once again to discover more about the analytic properties of $S_l(k)$. If we were to solve Eq. 1.33 at $k = k'$, which may be complex, then we have

$$\left[\frac{d^2}{dr^2} + (k')^2 - \frac{l(l+1)}{r^2} - 2U(r)\right]F_l(k'r) = 0 \tag{1.80}$$

with the asymptotic form

$$F_l(k'r) \underset{r\to\infty}{\sim} [e^{-i(k'r-\frac{1}{2}l\pi)} - S_l(k')e^{i(k'r-\frac{1}{2}l\pi)}], \tag{1.81}$$

to within an overall, complex, r-independent, but k' and l-dependent, constant. Furthermore, if we were to solve the equation a second time at the mirror image point, $k = -k'$ we should be solving the radial equation

$$\left[\frac{d^2}{dr^2} + (-k')^2 - \frac{l(l+1)}{r^2} - 2U(r)\right]F_l(-k'r) = 0 \tag{1.82}$$

which is identical to Eq. 1.80, with the asymptotic form

$$F_l(-k'r) \underset{r\to\infty}{\sim} [e^{-i(-k'r-\frac{1}{2}l\pi)} - S_l(-k')e^{i(-k'r-\frac{1}{2}l\pi)}] \tag{1.83}$$

Equation 1.83 can be rewritten by pulling $S_l(-k')$ out of the square brackets

$$F_l(-k'r) \underset{r\to\infty}{\sim} -S_l(-k')[e^{-i(k'r+\frac{1}{2}l\pi)} - S_l(-k')^{-1}e^{i(k'r+\frac{1}{2}l\pi)}].$$

In order to get the phase factors in the exponentials in the same form as in Eq. 1.81 we substitute $e^{-il\pi}e^{il\pi} = 1$ between $S_l(-k')$ and the square bracket and take the right-hand factor $e^{il\pi}$ into the square bracket to obtain

$$F_l(-k'r) \sim -S_l(-k')e^{-il\pi}[e^{-i(k'r-\frac{1}{2}l\pi)} - e^{il\pi}S_l(-k')^{-1}e^{+il\pi}e^{i(k'r-\frac{1}{2}l\pi)}] \tag{1.84}$$

Since $e^{2il\pi} = 1$, then comparison of Eq. 1.84 with Eq. 1.81 gives the relation between S and its inverse on the complex k-plane to be

$$S_l(k') = S_l(-k')^{-1}. \tag{1.85}$$

From Eq. 1.85 we see that if the S-matrix has a zero at $k' = -k$, then the inverse has a zero at $(+k)$! Since

$$S_l(k')^{-1} = \frac{1}{S_l(k')} = 0, \tag{1.86}$$

then we must have that $S_l(k')$ has a pole where its inverse has a zero. Returning to the discussion earlier in this section, we recall that bound states corresponded to zeros of $S_l(k)$ on the negative imaginary axis, which we now see corresponds to zeros of the inverse S-matrix on the positive imaginary axis, which according to Eq. 1.86 corresponds to *poles* of the S-matrix on the positive imaginary axis, see Fig. 10.

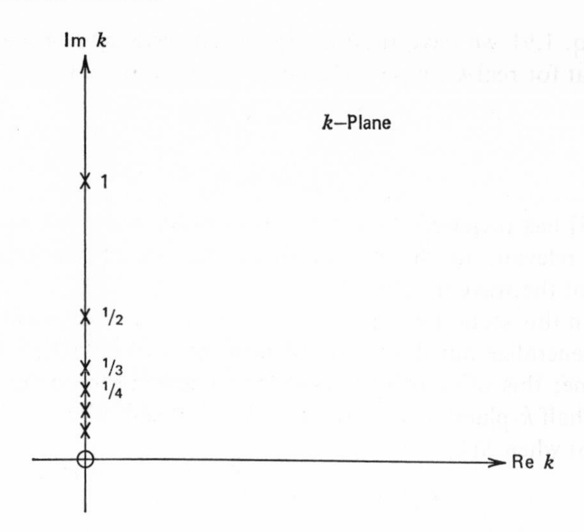

Figure 10 Location of the poles of the S-matrix on the complex k-plane for the hydrogen atom.

If we were to solve the complex conjugate of Eq. 1.33, for real l, we would be solving

$$\left(\frac{d^2}{dr^2} + (k)^{2*} - \frac{l(l+1)}{r} - 2U(r)\right) F_l^*(kr) = 0, \tag{1.87}$$

with the asymptotic form derived by taking the complex conjugate of Eq. 1.41, that is

$$F_l^*(kr) \underset{r\to\infty}{\sim} e^{i(k^*r - \frac{1}{2}l\pi)} - S_l^*(k)e^{-i(k^*r-\frac{1}{2}l\pi)}. \tag{1.88}$$

If we were to solve the radial equation at $k = k'^*$, then Eq. 1.88 becomes

$$F_l^*(k'^*r) \underset{r\to\infty}{\sim} [e^{i(k'r-\frac{1}{2}l\pi)} - S_l^*(k'^*)e^{-i(k'r-\frac{1}{2}l\pi)}]. \tag{1.89}$$

In Eq. 1.81 we see that we can factor out the S-matrix from the square brackets to obtain

$$F_l(k'r) \underset{r\to\infty}{\sim} -S_l(k')[e^{i(k'r-\frac{1}{2}l\pi)} - S_l(k')^{-1}e^{-i(k'r-\frac{1}{2}l\pi)}] \tag{1.90}$$

which upon comparison with Eq. 1.89 gives

$$S_l(k')^{-1} = S_l^*(k'^*), \tag{1.91}$$

because $(k'^*)^{2*} = (k'^*k'^*)^* = k'^2$, which means that we solved identically the same radial equation to get the seemingly different asymptotic forms (1.89) and (1.90).

From Eq. 1.91 we have the very important unitarity property of the S-matrix, that for real k', $k' = k'^* = k$,

$$S_l(k)^{-1} = S_l^*(k),$$

or

$$|S_l(k)|^2 = 1. \tag{1.92}$$

Newton [5] has reviewed the properties of radial wave functions and other quantities relevant to the partial wave analysis of scattering theory as functions of the wave number, k.

Earlier in this section we considered zeros on the negative imaginary axis. We now generalize our discussion to zeros of $S(k)$ anywhere on the lower-half k-plane; this of course corresponds to considering poles anywhere on the upper-half k-plane. If we write $k = k_1 - i\kappa$, κ real and > 0, k_1 real, then from (1.76) when $S(k) = 0$

$$F_l(kr) \underset{r\to\infty}{\sim} e^{-ik_1r}e^{-\kappa r}, \qquad \operatorname{Im} k < 0 \tag{1.93}$$

which describes a bound state, and F_l is normalizable. We proceed to prove that in fact the bound state zeros cannot lie anywhere in the lower-half k-plane! Premultiply Eq. 1.33 by $F_l^*(kr)$ and Eq. 1.87 by $F_l(kr)$ and subtract to obtain

$$F_l^*\left[\frac{d^2}{dr^2}F_l + k^2F_l\right] - F_l\left[\frac{d^2}{dr^2}F_l^* + k^{2*}F_l^*\right] = 0,$$

When this equation is integrated from zero to infinity the differential terms vanish because of the boundary conditions (1.93) and at the origin, leaving

$$(k^2 - k^{2*})\int_0^\infty drF_l^*(kr)F_l(kr) = 0, \tag{1.94}$$

which can only be true, for nontrivial radial functions, provided

$$k^2 = k^{2*}, \text{ that is, } \operatorname{Re} k \equiv 0, \tag{1.95}$$

see Fig. 10.

Equation 1.95 proves that the poles of the S-matrix in the upper-half k-plane are confined to the positive imaginary axis. Since we cannot make any similar statement about zeros in the upper-half k-plane, then in general the location of the *poles* over the entire k-plane are as drawn in Fig. 11.

From Eqs. 1.85 and 1.91 we have that

$$S_l(k) = S_l^*(-k^*), \tag{1.96}$$

so that if $S_l(k)$ has a zero at $k = k' = k_1 + ik_2$, then S_l^* has a zero at $-k'^* = -k_1 + ik_2$, the conjugate zero of S_l. Hence the symmetry of the poles and their conjugate poles on the lower-half k-plane drawn in Fig. 11.

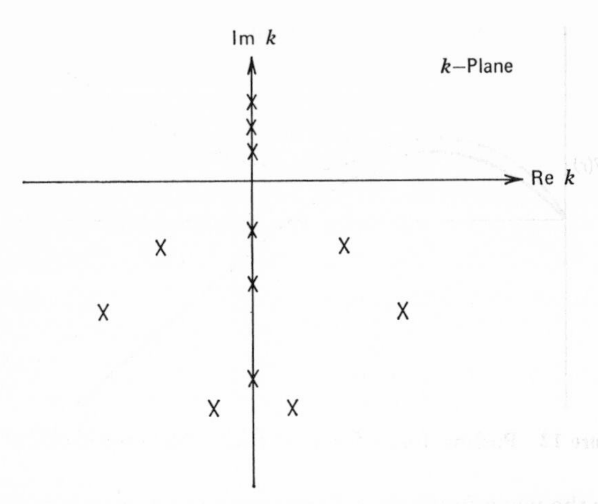

Figure 11 General location of the poles of the single channel S-matrix on the complex k-plane.

1.2.2. Virtual States

In Fig. 11 we have drawn some poles on the *negative* imaginary k-axis; we should like to know their physical significance, if any. As far as bound states are concerned we know that the more tightly bound a system, the farther up the imaginary axis is located the corresponding pole in $S_l(k)$. This is obvious for the Coulomb problem, where r^{-1} supports an infinite number of bound states, see Fig. 10. In contrast to this potential, it is well known that the neutron–proton interaction in the ^3S state supports only a single bound state, the deuteron. That is to say, there exists only *one* value of k^2 on the real negative axis of the k^2 plane for which the solution to the radial equation obeys the stationary state boundary conditions. At the threshold for scattering, $k^2 = 0$, the radial equation becomes (for s-waves)

$$\left[\frac{d^2}{dr^2} - 2U(r)\right]F(r) = 0. \tag{1.97}$$

In the asymptotic domain, where $U(r)$ can be neglected, the solution is a straight line

$$F(r) \underset{r\to\infty}{\sim} C_2 r + C_1 \tag{1.98}$$

where C_1 and C_2 are integration constants and the wave function has a shape as drawn in Fig. 12. At small r, where $|U(r)| \gg k^2$, the wave function will have shapes which are almost k^2 independent.

In neutron–proton scattering in the ^1S state, the short-range attractive two-nucleon potential is not strong enough to support even a single bound state, and the wave function at $k^2 = 0$ follows the shape given in Fig. 12 in the dashed curve [6]. In other words, the potential is not sufficiently attractive

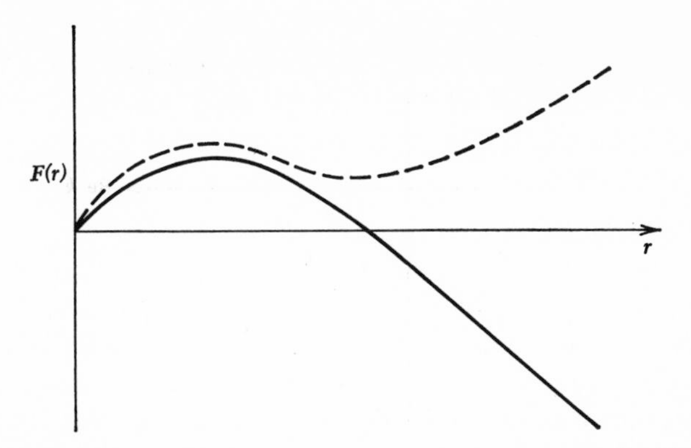

Figure 12 Possible forms for a wave function at threshold, $k^2 = 0$.

to bend over the wave function, as drawn for the continuous curve. A virtual state is one which would be bound if the interaction was more attractive; mathematically, we can define a virtual state as a zero of the S-matrix on the positive imaginary k-axis. The closer the zero is to the real k-axis, the closer will be the corresponding pole on the negative k^2-axis to the elastic scattering threshold and the greater will be the influence of the singularity on the S-matrix at, and beyond the threshold, $k^2 = 0$.

In summary, then, we have encountered scattering states, the real positive energy axis, $k^2 > 0$, bound states corresponding to poles on the negative real energy axis, $k^2 < 0$, and virtual states, see Fig. 13.

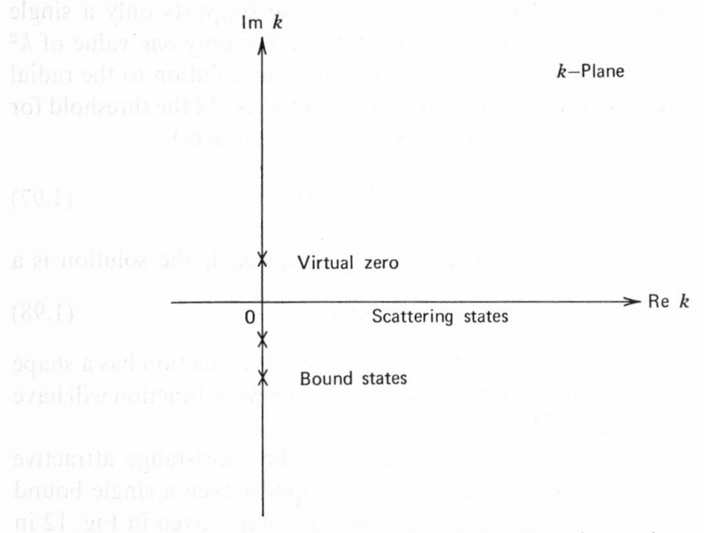

Figure 13 Summary of analytic properties of the S-matrix on the complex k-plane.

Problem 7. Given the attractive two-body potential to be $-U_0e^{-r/a}$, derive an expression for the scattering matrix and discuss the bound states of the system.

1.2.3. Decaying States

In the two previous subsections we have been concerned with the physical interpretation of the zeros of the S-matrix which may be found on the imaginary axis of the k-plane. The question still remains of the interpretation of zeros of S which may be located at complex $k = -k_0 + ik_1$, with k_0 and k_1 real. From Eq. 1.41 we see that at this complex zero of $S_l(k)$

$$F_l(kr) \underset{r\to\infty}{\sim} e^{-i(-k_0+ik_1)r} \underset{r\to\infty}{\sim} e^{ik_0r}e^{k_1r} \tag{1.99}$$

which will be an *outgoing* radial wave if $k_0 > 0$! That is to say, the solution of the radial equation at this energy for which $S(k) = 0$, does *not* contain any component of an ingoing wave.

From the quantum mechanical definition of probability current, the number of particles passing through the surface of a sphere of radius a per unit time is

$$N = -\frac{i\hbar}{2m}\left(\frac{\partial F}{\partial r}F^* - \frac{\partial F^*}{\partial r}F\right)4\pi a^2 = 4\pi a^2\hbar k_0 e^{2k_1a/m}$$

When we define

$$|\psi(r)|_a{}^2 = e^{2k_1a} \tag{1.101}$$

we have that

$$N = 4\pi a^2 \cdot |\psi(r)|_a{}^2 \cdot v,$$

where $v \equiv \hbar k_0/m$ is the velocity of the emerging particles. The time-dependent wave function is

$$\Psi(\mathbf{r}, t) = \psi(r)e^{-iEt/\hbar}$$

where

$$\begin{aligned} E &= \frac{\hbar^2}{2m}k^2 = \frac{\hbar^2}{2m}(-k_0 + ik_1)^2 = \frac{\hbar^2}{2m}(k_0{}^2 - k_1{}^2 - 2ik_0k_1) \\ &\equiv E_0 - i\frac{\Gamma}{2}, \quad \text{with} \quad \frac{\Gamma}{2} \equiv k_0k_1\frac{\hbar^2}{m}. \end{aligned} \tag{1.102}$$

Finally, the probability density is given by

$$|\Psi(\mathbf{r}, t)|^2 = e^{2k_1a}e^{-t/\tau}, \qquad \tau \equiv \frac{\hbar}{\Gamma}\text{ sec}, \tag{1.103}$$

having used Eq. 1.101. For $k_0 > 0$ and $k_1 > 0$, then $\Gamma > 0$ and the system described by the state function Ψ decays with a mean lifetime of τ secs. For virtual states, $k_0 = 0$ and so $\Gamma = 0$. States that are defined by Eq. 1.103 are known as quasi-stationary states. The existence of such states forms the basis of resonance theory. From Eq. 1.103 we see that the probability is not time independent, as a bound state system is, but rather it leaks away in time. For reasons to be given later, the quantity Γ is called the width of the state and is usually measured in electron volts. These decaying states of quantum mechanical systems are also known as Siegert states [7].

When we recall Eq. 1.86, and the subsequent discussion that if the S-matrix has a zero at $k = -k'$, then it has a pole at $k = k'$, we are led to introduce the term resonance pole—those poles which correspond to the zeros of $S(k)$ at complex k and which are identified with wave functions that are asymptotic to outgoing waves only.

Consider Eq. 1.75—s-wave potential scattering with $F(0) = 0$. Since this equation is a second-order ordinary differential equation, then there are two integration constants, the boundary condition at the origin determines one of them and the second can be chosen arbitrarily; for example, the first derivative at the origin, $F(0)'$, a real number, since the equation is homogeneous in F. With these two boundary conditions at $r = 0$, Eq. 1.75 can be integrated step by step out from the origin into the asymptotic region. For real $k^2 > 0$, the numerical values obtained for $F(r)$ must be of the form

$$F(r) \sim a \sin kr + b \cos kr, \tag{1.104}$$

while for real $k^2 < 0$

$$F(r) \sim a e^{|k|r} + b e^{-|k|r}, \qquad S \equiv \frac{-b}{a}. \tag{1.105}$$

In both these equations the integration constants a and b must be real, since the boundary conditions are real and the equation is real, consequently the solution is real.

From Eq. 1.105 we see that if there exists a value of k^2 such that $a = 0$, then

1. $F(r) \sim b e^{-k|r|}$, which is a bound state wave function.
2. $S = \infty$, a pole on the negative k^2-axis.

If we rewrite Eq. 1.104 as

$$F(r) \sim \left(-\frac{a}{2i} + \frac{b}{2}\right) e^{-ikr} + \left(\frac{a}{2i} + \frac{b}{2}\right) e^{ikr} \tag{1.106}$$

then we can only get outgoing waves provided

$$-\frac{a}{2i} + \frac{b}{2} = 0. \tag{1.107}$$

For real $k^2 > 0$, this is only possible provided $a = 0 = b$ and $F(r)$ is the trivial

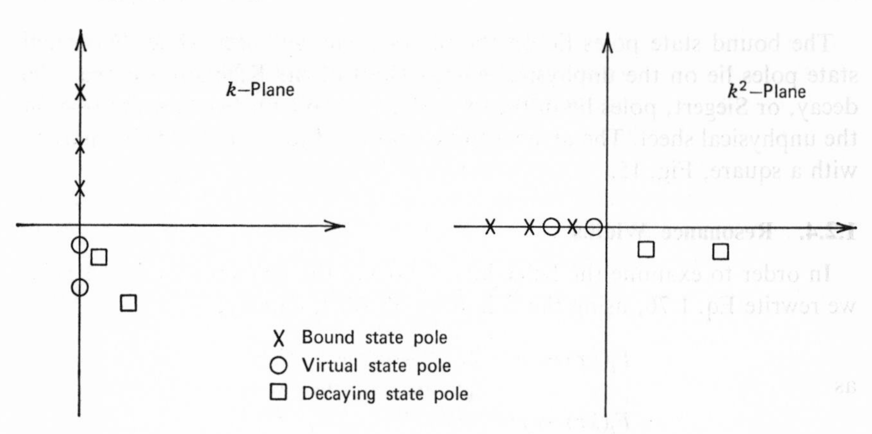

Figure 14 Location of poles of the single channel S-matrix on the complex k and k^2-planes.

solution. We therefore conclude that for real $k^2 > 0$ it is not possible to have a nontrivial decay solution to Eq. 1.75. However, for complex k^2, even with real starting conditions, $F(r)$ can be complex in the asymptotic region and it will be possible to satisfy (1.107) with nontrivial a and b. The above arguments have demonstrated that by mere inspection of the ordinary second-order differential equation that solutions, with outgoing radial waves only, can be found only for complex k^2.

In any scattering experiment we must have k^2 real and positive and so we can never have a system which is only decaying. Theoretically, if $S(k)$ has a pole in the vicinity of the physical energy axis, then we should expect such a singularity to have quite a dramatic influence on the physical quantities, phase shifts, and cross sections.

In Eq. 1.99 we have noted that we can only have an outgoing wave provided $k_0 > 0$ and in Eq. 1.103 we can only have positive lifetimes provided $k_1 > 0$. The zeros of $S(k)$ are then at $k = -k_0 + ik_1$ and the corresponding pole in $S(k)$ is at $k = k_0 - ik_1$, see Fig. 14.

From the simple discussion on poles and zeros we know that $S(k) \neq S(-k)$. However, $k_1{}^2 = (\pm k_1)^2$, so the two points on the k-plane, $+k_1$ and $-k_1$, are mapped onto a single point $k_1{}^2$ of the k^2-plane. That is to say, $S(k^2)$ is a double-valued function of k^2 with a square-root branch point at the origin of the k^2 plane. The usual convention for "raising" this duality is to take the values of $S(k)$ on the upper-half k-plane to be the physical sheet of the Riemann surface of $S(k^2)$ values, while the lower-half k-plane is taken to be the unphysical sheet. This means that the branch cut has been chosen to run from the branch point at $k^2 = 0$ along the real positive k^2-axis. The physical collision values of $S(k)$ are defined to be the limit as the real axis is approached from above on the physical sheet, see Fig. 15.

The bound state poles lie on the physical energy sheet while the virtual state poles lie on the unphysical energy sheet of the Riemann surface. The decay, or Siegert, poles lie in the lower-half k-plane and so they are also on the unphysical sheet. The arrow can be continued to the decay pole, marked with a square, Fig. 15.

1.2.4. Resonance Widths

In order to examine the behavior of $\delta_l(k)$ in the neighborhood of a pole, we rewrite Eq. 1.76, using the definition of $S_l(k)$, namely,

$$F_l(kr) \sim e^{-i(kr-l\pi/2)} - e^{2i\delta_l}e^{i(kr-l\pi/2)},$$

as

$$F_l(kr) \sim e^{-i(kr-l\pi/2+\delta_l)} - e^{i(kr-l\pi/2+\delta_l)},$$

since F_l is arbitrary to within an r-independent constant, and which can be written as

$$F_l(E, r) \sim A_l(E)e^{-ikr} - A_l^*(E)e^{ikr} \tag{1.108}$$

where we have introduced the complex functions of complex E

$$A_l(E) = i^l e^{-i\delta_l(E)}, \qquad A_l^*(E) = (-i)^l e^{i\delta_l(E)}. \tag{1.109}$$

For a decay state, we must have

$$A_l(E) = 0 \quad \text{at} \quad E = E_r - i\frac{\Gamma}{2},$$

where E_r and Γ are real and positive, since

$$E = \frac{\hbar^2}{2m}k^2 = \frac{\hbar^2}{2m}(-k_0 + ik_1)^2 \equiv E_r - i\frac{\Gamma}{2}. \tag{1.110}$$

As a matter of notation we should have a suffix l on both E_r and Γ, since we

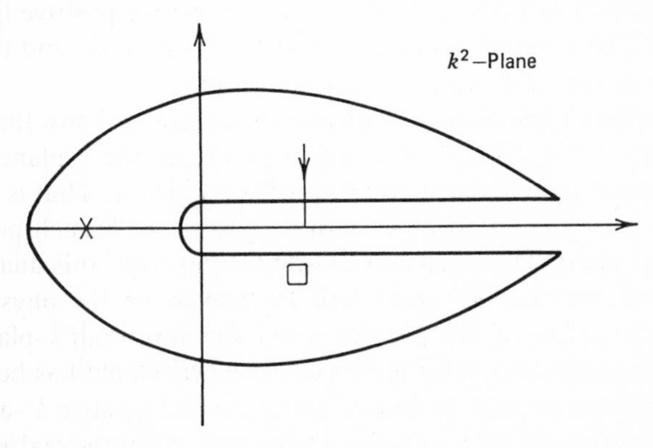

Figure 15 Schematic of the branch cut for the elastic threshold along the positive energy axis.

are investigating the existence of a pole in a particular partial wave l. Using Taylor series, we can expand $A_l(E)$ in the neighborhood of its zero

$$A_l(E) \approx A_l\left(E_r - i\frac{\Gamma}{2}\right) + \left\{E - \left(E_r - i\frac{\Gamma}{2}\right)\right\}a_l + \cdots$$

where a_l is the first derivative of $A_l(E)$ evaluated at the zero. Since the first term is zero, then

$$A_l(E) \approx \left\{E - \left(E_r - i\frac{\Gamma}{2}\right)\right\}a_l + \cdots \tag{1.111}$$

and

$$A_l^*(E) \approx \left\{E - \left(E_r + i\frac{\Gamma}{2}\right)\right\}a_l^* + \cdots$$

Substituting these results into Eq. 1.108 gives

$$\begin{aligned} F_l(E, r) &\sim \left\{E - \left(E_r - i\frac{\Gamma}{2}\right)\right\}a_l e^{-ikr} - \left\{E - \left(E_r + i\frac{\Gamma}{2}\right)\right\}a_l^* e^{ikr} \\ &\sim i\Gamma a_l^* e^{ikr}, \qquad \text{at} \quad E = E_r - i\frac{\Gamma}{2}. \end{aligned} \tag{1.112}$$

We note that the position of the pole is defined by its distance along the positive energy axis, E_r, as well as distance below this axis, $\Gamma/2$, into the lower-half energy plane. When we compare this form of $F_l(E, r)$, in the *neighborhood* of the pole in the S-matrix, with the form we started this section with, we get

$$\begin{aligned} e^{2i\delta_l(E)} &= e^{i\pi l}\frac{a_l^*}{a_l}\frac{E - E_r - i(\Gamma/2)}{E - E_r + i(\Gamma/2)} \\ &\equiv e^{2i\delta_l^{\text{pot}}}\left\{1 - i\frac{\Gamma}{E - E_r + i(\Gamma/2)}\right\} \end{aligned} \tag{1.113}$$

where we have defined the so-called potential, that is, nonresonant, part of the phase shift by

$$e^{2i\delta_l^{\text{pot}}} \equiv e^{i\pi l}\frac{a_l^*}{a_l}. \tag{1.114}$$

We see that for $|E - E_r| \gg \Gamma$, the second term in the braces of Eq. 1.113 can be neglected and $\delta_l \rightarrow \delta_l^{\text{pot}}$. In other words, in the energy region far from the influence of the zero, the phase shift δ_l is a smooth function of E, given by Eq. 1.114. The phase shift, δ_l^{pot}, is also called the background phase shift. The factor

$$\left\{1 - i\frac{\Gamma}{E - E_r + i(\Gamma/2)}\right\} = \frac{E - E_r - i(\Gamma/2)}{E - E_r + i(\Gamma/2)}$$

gives the rapid variation of $\delta_l(E)$ in the vicinity of the zero in $A_l(E)$. This part of $\delta_l(E)$ is called the resonant part of the phase shift, and we define

$$e^{2i\delta_l^{\text{res}}} \equiv \frac{E - E_r - i(\Gamma/2)}{E - E_r + i(\Gamma/2)} \equiv \frac{\cos\delta_l^{\text{res}} + i\sin\delta_l^{\text{res}}}{\cos\delta_l^{\text{res}} - i\sin\delta_l^{\text{res}}}. \tag{1.115}$$

We see now that we have separated the full partial wave phase shift into two components

$$e^{2i\delta_l} = e^{(2i\delta_l^{\text{pot}} + 2i\delta_l^{\text{res}})}; \qquad \delta_l = \delta_l^{\text{pot}} + \delta_l^{\text{res}}. \tag{1.116}$$

From Eq. 1.115 we have that

$$\frac{E - E_r}{-\Gamma/2} = \frac{\cos\delta_l^{\text{res}}}{\sin\delta_l^{\text{res}}}$$

hence

$$\cot\delta_l^{\text{res}}(E) = \frac{E_r - E}{\Gamma/2} \quad \text{and} \quad \delta_l^{\text{res}}(E) = -\cot^{-1}\frac{E - E_r}{\Gamma/2} \tag{1.117}$$

In order to examine the effect on δ_l^{res}, as E traverses a pole, consider

$$\tan\delta_l^{\text{res}} = \frac{\Gamma}{2}\Big/(E_r - E).$$

For $E < E_r$, $\tan\delta_l^{\text{res}}$ is positive; for $E > E_r$ we see that $\tan\delta_l^{\text{res}}(E)$ is negative. Therefore, for $\tan\delta_l^{\text{res}}$ to change sign as E goes through E_r, the phase shift itself δ_l^{res} must go through an odd multiple of $\pi/2$. In Fig. 16 we recall for the reader the graph of $\tan\theta$ versus θ. The crosses mark two points equal in magnitude but opposite in sign; as E traverses E_r, $\tan\delta_l^{\text{res}}$ must change from the positive cross to the negative cross. In other words, δ_l^{res} will be increased

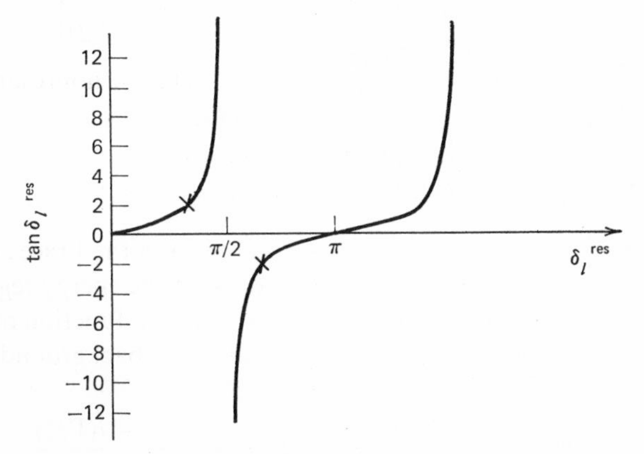

Figure 16 Plot of the tangent of the resonant part of the phase shift.

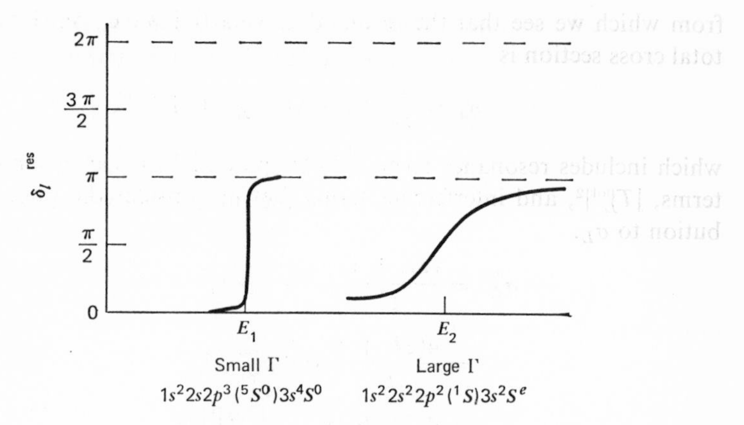

Figure 17 Variation of the resonant part of the phase shift with energy for $e - N^+$.

from 0 on the low-energy to π radians on the high-energy side of a Siegert pole. The pole coincides with δ_l^{res} equal to an odd multiple of $\pi/2$. Thus, the full phase-shift, δ_l, will be incremented by π radians each time a pole is traversed as the energy is increased.

In Fig. 17 we show how the resonant part of the phase shift varies over the energy region of the pole with small width Γ (i.e., close to the real energy axis) and for poles quite distant from the real k^2 axis (i.e., large Γ). The smaller Γ, the more rapid increase in δ_l^{res}. The two examples presented are found in the scattering of electrons by singly charged nitrogen ions and are due to autionized states (more later) with the configurations given in the figure.

The separation of the phase shift into potential and resonant contributions corresponds to an equivalent separation of the interaction potential which produces the phase shift. In the single channel problem the potential is thought of as having two components, usually a long-range repulsive component which would give δ_l^{pot} if we solved the radial equation with this component alone, and a short-range attractive component of the potential which provides the mechanism for generating the resonance.

We have defined the T-matrix for potential scattering by

$$\begin{aligned} T_l(E) &= e^{2i\delta_l} - 1 = e^{2i\delta_l} - e^{2i\delta_l{}^{\mathrm{pot}}} + e^{2i\delta_l{}^{\mathrm{pot}}} - 1 \\ &= e^{2i\delta_l{}^{\mathrm{pot}}}(e^{2i\delta_l{}^{\mathrm{res}}} - 1) + (e^{2i\delta_l{}^{\mathrm{pot}}} - 1) \\ &\equiv T_l^{\mathrm{res}} + T_l^{\mathrm{pot}} \end{aligned} \tag{1.118}$$

and the total cross section has been derived in units of πa_0^2 to be, see Eq. 1.43,

$$\sigma(E) = \frac{1}{k^2} \sum_l (2l + 1)\,|T_l|^2, \tag{1.119}$$

from which we see that the resonant Lth partial wave contribution to the total cross section is

$$\sigma_L = \frac{1}{k^2}(2L+1)\,|T_L^{\text{res}} + T_L^{\text{pot}}|^2, \tag{1.119a}$$

which includes resonance terms, $|T_L^{\text{res}}|^2$, potential or background scattering terms, $|T_L^{\text{pot}}|^2$, and interference terms. Let us consider the resonant contribution to σ_L.

$$\begin{aligned}\sigma_L^{\text{res}} &= \frac{(2L+1)}{k^2}\,|e^{2i\delta_L^{\text{res}}} - 1|^2 \\ &= \frac{4(2L+1)}{k^2}\sin^2\delta_L^{\text{res}} \\ &= \frac{4(2L+1)}{k^2}\cdot\frac{(\Gamma/2)^2}{(E-E_r)^2 + (\Gamma/2)^2},\end{aligned} \tag{1.120}$$

having used Eq. 1.117.

Equation 1.120 is known as the Breit-Wigner formula for resonant cross sections and has the shape drawn in Fig. 18, where we have presented the actual results for electrons scattered by N^+ when the total spin of the system is $\frac{1}{2}$ and the total orbital angular momentum is zero. The energy-value E_r is defined to be the position of the resonance of a cross section, that is, the energy at which $\delta_L^{\text{res}}(E) = (2n+1)(\pi/2)$ and *not* where the full phase shift $\delta_L = (2n+1)(\pi/2)$. At $E = E_r \pm \Gamma/2$ we note from Eq. 1.120 that σ_L^{res} is equal to half its maximum value. This accounts for the name of Γ—the "width" of the resonant part of the cross section at half-maximum.

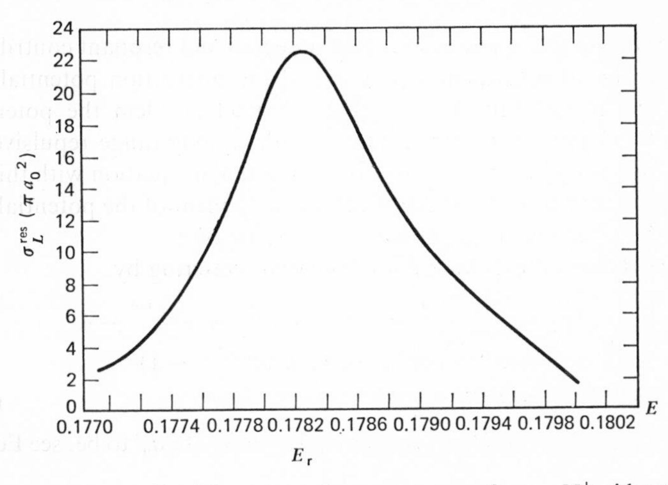

Figure 18 Resonant part of the cross section versus energy for $e - N^+$ with total quantum numbers $^2S^e$.

As demonstrated by Eq. 1.119, the resonant part of the partial wave cross section is superimposed on the background cross section for that partial wave, as well as the contributions from all other partial waves $l \neq L$. Whether an experiment will detect such a resonance depends on the extent of the interference between background and resonance; it will depend on Γ—broad resonances are normally difficult to observe—and it will depend on the other σ_l. It should be emphasized that Siegert poles are associated with a particular partial wave. Since Γ is also a measure of how far the Siegert pole in the S-matrix is from the real energy axis, see Eq. 1.112, we now see the meaning of the qualitative statement, that distant Siegert poles will have little effect on $\sigma(E)$, which can only be observed for real, positive E.

1.2.5. Resonance Shapes

Now that we have introduced the concept of a resonance width, Γ, and position, E_r, which are related to the complex and real parts respectively of the position of the Siegert pole, see Eq. 1.110, we can examine the influence of the interference term in Eq. 1.119.

$$
\begin{aligned}
T_l^{\mathrm{res}*} T_l^{\mathrm{pot}} + T_l^{\mathrm{res}} T_l^{\mathrm{pot}*} \\
&= e^{-i\delta_l^{\mathrm{pot}}} e^{-i\delta_l^{\mathrm{res}}} (e^{-i\delta_l^{\mathrm{res}}} - e^{i\delta_l^{\mathrm{res}}})(e^{i\delta_l^{\mathrm{pot}}} - e^{-i\delta_l^{\mathrm{pot}}}) \\
&\quad + e^{i\delta_l^{\mathrm{pot}}} e^{i\delta_l^{\mathrm{res}}} (e^{i\delta_l^{\mathrm{res}}} - e^{-i\delta_l^{\mathrm{res}}})(e^{-i\delta_l^{\mathrm{pot}}} - e^{i\delta_l^{\mathrm{pot}}}) \\
&= 2 \sin \delta_l^{\mathrm{res}} 2 \sin \delta_l^{\mathrm{pot}} [e^{-i(\delta_l^{\mathrm{pot}} + \delta_l^{\mathrm{res}})} + e^{i(\delta_l^{\mathrm{pot}} + \delta_l^{\mathrm{res}})}]
\end{aligned}
$$

Consequently we can write the interference contribution to the Lth partial wave cross section

$$\sigma_L^{\mathrm{int}} \propto \sin \delta_L^{\mathrm{res}} \sin \delta_L^{\mathrm{pot}} \cos (\delta_L^{\mathrm{res}} + \delta_L^{\mathrm{pot}}). \tag{1.121}$$

It is this contribution to the total cross section which is responsible for the asymmetry observed in the experimental cross sections.

In Fig. 19 we have drawn the full phase shift and partial wave cross section, as a function of E, for a variety of values of the background phase shift.

a. Away from the resonance, there is no background scattering, $\delta_l^{\mathrm{pot}} = 0$, and so the cross section is given completely by the Breit-Wigner formula.

b. Away from the resonance, there is a small positive background phase shift, δ_l^{pot}, and in this energy region the partial wave cross section is wholly potential scattering, namely,

$$\sigma_L \approx \sigma_L^{\mathrm{pot}} = \frac{4(2L+1)}{k^2} \sin^2 \delta_L^{\mathrm{pot}} = \frac{4(2L+1)}{k^2} \sin^2 [\delta_L^{\mathrm{pot}} + m\pi], \tag{1.122}$$

for $m = 0, 1, 2, \ldots$.

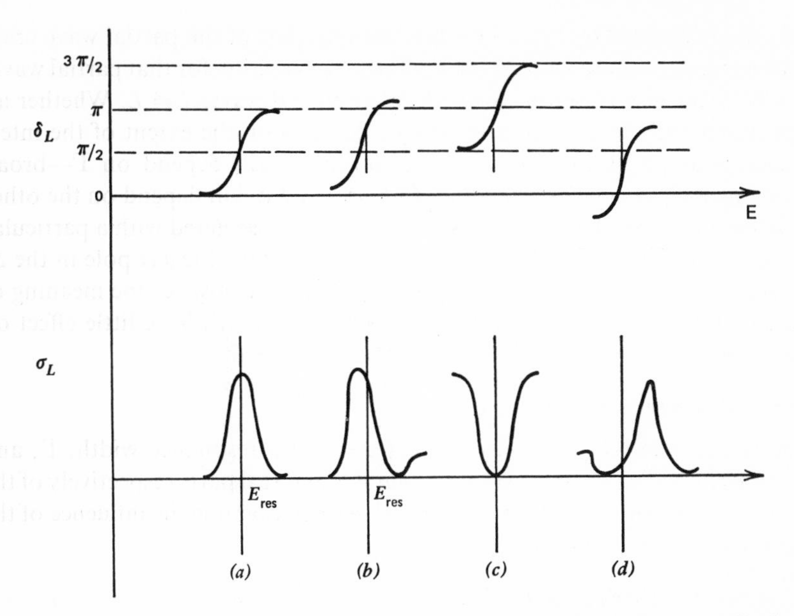

Figure 19 Various resonance profiles, σ_L, and corresponding full phase shift, δ_L.

We have seen that as tan δ_L^{res} traverses a resonance it changes from 0+, where the phase shift $\delta_L = \delta_L^{\text{pot}} + m\pi$, to 0−, where the full partial wave phase shift will have been increased by π radians, that is, $\delta_L = \delta_L^{\text{pot}} + (m+1)\pi$. Consequently, as the phase shift traversed a sequence of resonances the phase shift would be increased by π radians at every Siegert pole and we might plot the continuous curve as in Fig. 20. In this figure we present actual

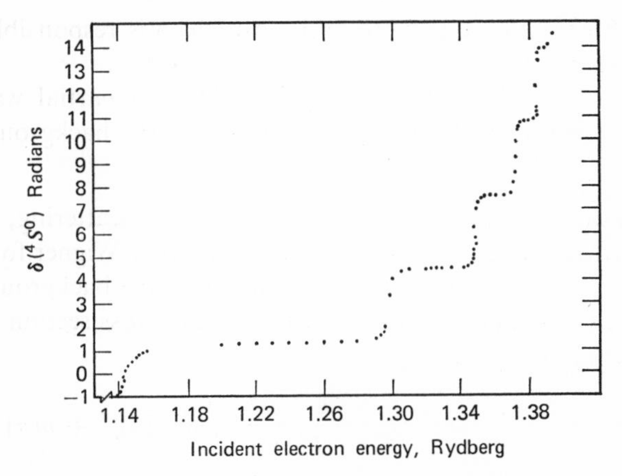

Figure 20 Phase shift for electrons on N^+ $(1s^22s2p^3(^3S^0)ns\,^4S^0$ series).

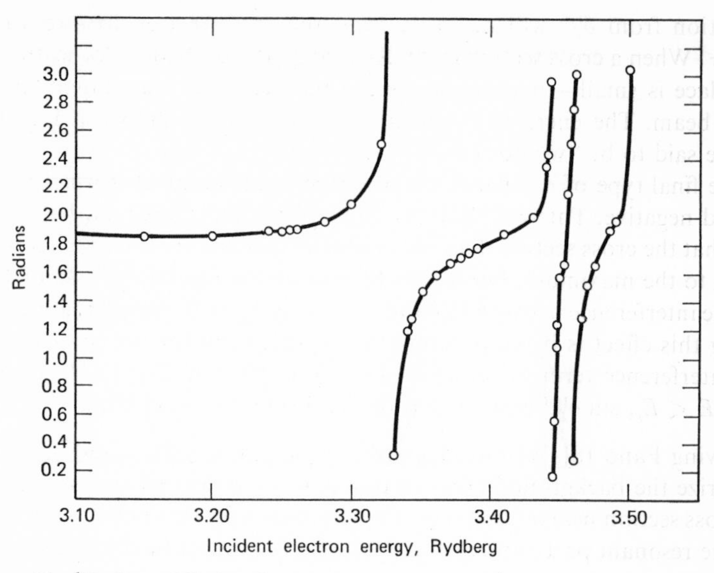

Figure 21 ^1P phase shift for electrons scattered by He^+ in the six-state approximation, that is, including the $1s$-$2s$-$2p$-$3s$-$3p$-$3d$ states.

calculated values of $\delta_{L=0}$ for electrons on N^+. We notice how the rate of increase of δ with E increases as we go from resonance to resonance at higher energies. This is because Γ is decreasing. If there are many resonances, such a continuous curve is impractical and so an odd multiple of π is subtracted to give figures which are often plotted like Fig. 21, where we have plotted the $\delta(^1P)$ phase shift for electrons on He^+.

In this case of a small positive δ_L^{pot} we know that σ_L will depend on δ_L to the extent that σ_L will reach its maximum *before* E_r. Since $\sigma_L \propto \sin^2 \delta_L$, the partial wave cross section will fall to zero when δ_L goes through $m\pi$ on the high energy side of E_r. These qualitative remarks are substantiated by Eq. 1.122: throughout the whole energy region, δ_L^{pot} is usually a smoothly varying function of E. To examine the qualitative effect of the interference terms as given by Eq. 1.121, we note that while $\sin \delta_L^{\text{res}}$ is positive on the low-energy side of E_r it is negative on the high-energy side of E_r; the former ensures that the interference term adds to $(\sigma_L^{\text{pot}} + \sigma_L^{\text{res}})$ while the latter ensures that it subtracts.

c. In this case we take the background phase shift to be $\pi/2$, and from Eq. 1.122 we shall have that

$$\sigma_L^{\text{pot}} = \frac{4(2L+1)}{k^2}, \tag{1.123}$$

which is the maximum value that even σ_L can have. Consequently, any contribution from δ_L^{res} will be to decrease the cross section to zero when $\delta_L = m\pi$! When a cross section is small—that is, the probability of scattering taking place is small—then the target can be said to be transparent to the incident beam. The energies at which the cross sections drawn in Fig. 19c occur are said to be "windows."

d. The final type of resonance shape which is encountered is when δ_L^{pot} is small and negative. This will delay δ_L from going though $\pi/2$ until after E_r. We see that the cross section does *not* gradually increase from its background value up to the maximum, but it goes to zero on the low-energy side of E_{res} due to the interference between δ_L^{pot} and δ_L^{res} giving $\delta_L = 0$. An alternative way of seeing this effect is by considering the explicit form for the contribution of the interference term to the cross section as given in Eq. 1.121. We see that for $E < E_r$, $\sin \delta_L^{\text{pot}}$ ensures that this contribution is negative.

Following Fano [8], we introduce two *shape* parameters, ε and q, which characterize the background cross section and the maximum and minimum of the cross section near a resonance. The expression for the transition matrix, T, for the resonant partial wave L given in Eq. 1.118 can be rewritten in the form

$$\begin{aligned} T_L &= e^{2i\delta_L} - 1 = e^{i\delta_L}[e^{i\delta_L} - e^{-i\delta_L}] \\ &= 2ie^{i\delta_L} \sin \delta_L = 2ie^{i\delta_L} \sin (\delta_L^{\text{res}} + \delta_L^{\text{pot}}) \\ &= 2ie^{i\delta_L}[\sin \delta_L^{\text{res}} \cos \delta_L^{\text{pot}} + \cos \delta_L^{\text{res}} \sin \delta_L^{\text{pot}}] \\ &= 2ie^{i\delta_L} \sin \delta_L^{\text{pot}}[\cot \delta_L^{\text{pot}} + \cot \delta_L^{\text{res}}] \sin \delta_L^{\text{res}} \end{aligned} \tag{1.124}$$

We can define the shape parameters by

$$q \equiv -\cot \delta_L^{\text{pot}} \quad \text{and} \quad \varepsilon \equiv -\cot \delta_L^{\text{res}} = \frac{E - E_r}{\Gamma/2}, \tag{1.125}$$

see Eq. 1.117 for this latter result. The partial wave cross section can be written as, see Eq. 1.119,

$$\sigma_L \propto |T_L|^2 = 4 \sin^2 \delta_L^{\text{pot}} \frac{(\varepsilon + q)^2}{1 + \varepsilon^2}, \tag{1.126}$$

since

$$1 + \varepsilon^2 = 1 + \frac{\cos^2 \delta_L^{\text{res}}}{\sin^2 \delta_L^{\text{res}}} = \sin^{-2} \delta_L^{\text{res}}.$$

We note that $\sin^2 \delta_L^{\text{pot}}$ is directly related to the cross section for potential scattering, σ_L^{pot}. Furthermore, the total cross section

$$\sigma = \sum_{l=0}^{\infty} \sigma^l$$

will equal σ_L plus contributions from the remaining nonresonant partial waves (provided there is indeed only one resonant partial wave), $\bar{\sigma}$ say. The maximum and minimum of $\sigma(E)$ can be found immediately by differentiating with respect to ε, the formula

$$\sigma(\varepsilon) = \bar{\sigma} + \frac{4}{k^2}(2L+1)\sin^2\delta_L^{\text{pot}}\frac{(\varepsilon+q)^2}{1+\varepsilon^2} \tag{1.127}$$

The maximum is at $\varepsilon = q^{-1}$ and the minimum is at $\varepsilon = -q$. That is to say, neither extremes occur at E_r, where $\varepsilon = 0$.

It must be emphasized that in the above we have merely been examining the consequences of conjectures. That is to say, we have assumed the existence of poles off the real energy axis and then developed a formalism to describe and interpret data *on* that axis. So far we have said next to nothing about the mechanism which can generate a pole; in other words, we have not investigated at all the types of potentials which can produce Siegert poles. For bound states, we have merely said that $U(r)$ should be sufficiently attractive.

Fano and Cooper [9] have written Eq. 1.127 as

$$\sigma(\varepsilon) = \sigma_b + \sigma_a\frac{(q+\varepsilon)^2}{(1+\varepsilon^2)}, \tag{1.128}$$

where q, Γ, σ_a, and σ_b are regarded as practically energy independent in the proximity of a resonance.

At this point we introduce the term "autoionization," which can be observed in photon absorption spectra as *asymmetric* peaks, like those of Fig. 19, superimposed on a smooth background, at energies beyond the first ionization threshold. This phenomenon was interpreted as follows: in absorbing the photon, the atomic system undergoes a transition to a quasi-stationary state which subsequently releases an electron in a radiationless transition. Thus the autoionizing state is a decaying state and therefore corresponds to the Siegert pole of the S-matrix. Consequently, we can say that experimental observations of autoionization and resonance scattering are two ways of detecting the positions of the Siegert poles. The interrelationship between autoionization and resonance scattering theory has been brought out by Fano [8], who showed that the profiles (shapes) of absorption lines in the ionization continuum of atomic spectra are represented by Eq. 1.128.

The cross section obviously has a minimum at $\varepsilon = -q$, where it reduces to σ_b. As we have already said, this minimum appears in the spectra as a "transmission window." Far from the resonance, $\varepsilon \sim \pm\infty$, the value of the cross section is $(\sigma_a + \sigma_b)$. Consequently, at the point of minimum, the photon absorption is reduced in a ratio

$$\frac{\sigma_a}{\sigma_a + \sigma_b} \equiv \rho^2 \tag{1.129}$$

with respect to its value $\sigma(\pm\infty)$. This ratio is an important measurable index. The ratio ρ^2 is called a squared *correlation coefficient* by Fano and Cooper.

1.2.6. Resonances in Cross Sections

Let us summarize what we have said so far about resonances.

Experimentally, resonances are said to have occurred in scattering if the cross sections exhibit sharp maxima and minima. We have seen that the most succinct mathematical definition of a resonance, due to Siegert, is a pole of the S-matrix, located at such an energy that the Schrödinger equation has a solution with outgoing waves in all channels. The Siegert (resonance) poles of the S-matrix occur at complex energies, k_s^2; the imaginary part determining the width of the resonance, or the decay constant (lifetime) of the compound eigenstate of the dynamical system consisting of target and projectile. Some of these eigenstates, with a large width (i.e., short life) may be so far away from the physical energy axis that the broad resonances which they generate may not be recognized as resonances from the observed scattering. The various resonance theories can be thought of as different approximate methods for finding the Siegert poles.

The question as to whether all structures in cross sections should be ascribed to resonances has been answered by Gerjuoy [10] with a resounding *No*! Gerjuoy has pointed out that diffraction effects, which are in no way related to resonances, can cause fluctuations in the cross sections as functions of energy and angle. Other nonresonant effects which can account for cross-section rises and falls have been listed by Ford and Wheeler [11] and include essentially classical (i.e., nonquantum) effects such as rainbow scattering, glory scattering, and orbiting. Usually, the nonresonant phenomena can be distinguished from resonances due to the latter being associated with a single partial wave, while the former tends to result from the cooperative contributions from many partial waves.

As an example of the change of resonance profile with angle we consider scattering in which only s- and p-waves are important and the s-wave is resonant. The scattered intensity is given by Eq. 1.42 to be

$$\sigma(\theta) \approx \frac{1}{4k^2}\,|[e^{2i(\delta_s+\delta_r)} - 1] + 3\cos\theta[e^{2i\delta_p} - 1]|^2 \tag{1.130}$$

where δ_s and δ_p are the potential phase shifts and δ_r is the resonant part of the s-wave phase shift. We have

$$\begin{aligned}\sigma(\theta) &\approx \frac{1}{4k^2}\,[|T_s|^2 + 9\cos^2\theta\,|T_p|^2 + 3\cos\theta\{T_s{}^*T_p + T_sT_p{}^*\}]\\ &= \frac{1}{4k^2}\Bigg[4\sin^2\delta_s\frac{(\varepsilon+q)^2}{1+\varepsilon^2} + 9\cos^2\theta\cdot 4\sin^2\delta_p\\ &\quad + 3\cos\theta\cdot 4\sin\delta_0\sin\delta_p\cos(\delta_0+\delta_p)\Bigg],\end{aligned} \tag{1.131}$$

where $\delta_0 = \delta_s + \delta_r$.

1.3. STATIC FIELD APPROXIMATION

Until automatic digital computers become generally available, about 1956, most calculations on cross sections were performed in the Born approximation [12], or in the so-called static field approximation, which assumed that the projectile moved through a potential generated by the target without taking into consideration that there would be a disturbance at the target itself by the projectile. It will be shown below that such assumptions lead to the now familiar equation

$$\left[\frac{d^2}{dr^2} + k^2 - \frac{l(l+1)}{r^2} - U(r)\right]F_l(kr) = 0. \tag{1.132}$$

The solutions to this equation are interpreted as describing the radial motion of the projectile relative to the target.

In Section 1.4 it will be shown that at the origin, in order that the solution be physically meaningful, the boundary condition is

$$F_l(r) \underset{r\to 0}{\sim} r^{l+1}, \tag{1.133}$$

while in the asymptotic region (except for Coulomb potentials) for $k^2 > 0$ we have shown that

$$F_l(kr) \underset{r\to\infty}{\sim} kr[A_l(k)j_l(kr) - B_l(k)n_l(kr)] \tag{1.134}$$

In this section we describe a variety of collision processes which can be investigated using Eq. 1.132.

1.3.1. Electron and Positron Scattering by H-Atoms

The scattering of a projectile by a hydrogen atom is a three-body problem of the type posed in Problem 1. The Schrödinger equation describing the motion of the impinging particle and the bound atomic electron relative to the infinitely heavy proton is (having separated out the motion of the center-of-mass)

$$\left[\frac{-\hbar^2}{2m}\nabla_1{}^2 + \frac{-\hbar^2}{2m}\nabla_2{}^2 - \frac{e^2}{r_1} \mp \frac{e^2}{r_2} \pm \frac{e^2}{|r_1 - r_2|} - E\right]\Psi(\mathbf{r}_1, \mathbf{r}_2) = 0, \tag{1.135}$$

the upper signs are for incident electrons.

If the atomic electron, coordinate $\mathbf{r}_1$ say, is in its $1s$ ground state initially and remains in that state throughout the collision, then we can make the approximate separation of variables:

$$\Psi(\mathbf{r}_1, \mathbf{r}_2) \approx \psi_{1s}(\mathbf{r}_1)F(\mathbf{r}_2), \tag{1.136}$$

where ψ_{1s} satisfies the Schrödinger equation for the hydrogen atom

$$\left[\frac{-\hbar^2}{2m}\nabla_1^{\,2} - \frac{e^2}{r_1} - E_{1s}\right]\psi_{1s}(\mathbf{r}_1) = 0.$$

When Eq. 1.136 is substituted into (1.135) and the result premultiplied by the orthonormal function ψ_{1s}^* and then integrated over $\mathbf{r}$ we obtain the three-dimensional partial differential equation,

$$\{\nabla_2^{\,2} + k^2 - U(r_2)\}F(\mathbf{r}_2) = 0, \tag{1.137}$$

which is interpreted as describing the motion of the projectile relative to the target.

Problem 8. Show that

$$U(r) = \mp 2\left(1 + \frac{1}{r}\right)e^{-2r} \qquad \text{and} \qquad k^2 = 2m(E - E_{1s})/\hbar^2,$$

where the minus (plus) sign is for incident electrons (positrons).

The unknown coefficient function, $F(\mathbf{r})$, can be expanded in terms of Legendre polynomials which reduces the three-dimensional partial differential equation (1.137) to an infinite number of decoupled radial equations of the form of Eq. 1.132.

A technique for the numerical integration of this equation is given later. At this point we discuss qualitative features of the solution. To begin with, we know that a second-order ordinary differential equation has only two integration constants. Furthermore, we are only interested in solutions which vanish at the origin; this will determine one of the pair of integration constants. Equation 1.132 is homogeneous in F_l, consequently once we have found a solution we can multiply it by any (r-independent) constant and it will remain a solution. In other words, the normalization of F_l is arbitrary; and the second integration constant can remain arbitrary. For example, we could choose the slope at the origin to be anything we please. In this way, we see in Section 1.4 that we can establish a procedure for integrating the differential equation over a small interval of the independent variable $\Delta r = H$. The same procedure can be used repetitively out from the origin to values of r large enough that the potential $U(r)$ can be neglected, for $r \geq r_A$ say. At such distances the differential equation being solved is simply Bessel's equation with the solution given by Eq. 1.134.

At $r = r_A$, our algorithm for solving the radial equation yields numbers for $F_l(r_A)$ and $(dF_l/dr)|_{r_A}$. The functions $j_l(kr_A)$ and $n_l(kr_A)$ can be obtained, for example, from tables. Therefore, the unknown coefficients $A_l(k)$ and $B_l(k)$ can be determined from the pair of equations

$$\frac{F_l(r_A)}{kr_A} = A_l(k)j_l(kr_A) - B_l(k)n_l(kr_A)$$
$$\left[\frac{d}{dr}\left(\frac{F_l}{kr}\right)\right]_{r_A} = A_l(k)\left[\frac{dj_l}{dr}\right]_{r_A} - B_l(k)\left[\frac{dn_l}{dr}\right]_{r_A} \tag{1.138}$$

for each value of the energy k^2. The phase shift is given by Eq. 1.36, that is,

$$\tan \delta_l(k) = \frac{B_l(k)}{A_l(k)} \tag{1.139}$$

and the scattering intensity and elastic cross section can be computed from the formulas that we have established already.

1.3.2. Collisions of 100 keV Electrons with Au and Ag Atoms

An application of the methods just discussed to the scattering of electrons by Au and Ag atoms has a use in the phase contrast microscope. We shall be interested in knowing the difference in phase between the forward scattered wave, $\theta = 0$, and the incident wave. We have shown that the outgoing elastically scattered wave is given by, see Eq. 1.38a,

$$\frac{e^{ikr}}{r} \cdot \frac{1}{2ik} \sum_{l=0} (2l + 1)(e^{2i\delta_l} - 1)P_l(\cos\theta), \tag{1.140}$$

which we can write in the following form for the forward part of the scattered wave as

$$\frac{e^{ikz+i\chi}}{z} = (\cos\chi + i\sin\chi)\frac{e^{ikz}}{z} \tag{1.141}$$

where the phase angle χ is the quantity we are to calculate. Expanding $e^{2i\delta_l}$ into its real and imaginary components gives, for $\theta = 0$,

$$\cos\chi = \frac{1}{2k} \cdot \sum_{l=0}^{\infty} (2l + 1)\sin 2\delta_l$$
$$\sin\chi = \frac{1}{2k} \sum_{l=0}^{\infty} (2l + 1)(1 - \cos 2\delta_l). \tag{1.142}$$

Since gold and silver are many-electron systems, then a detailed quantum mechanical treatment of the scattering of electrons from these systems is very complicated. In this section we introduce a very simple model which assumes that the potential energy between the projectile and target can be represented by a screened Coulomb potential of the form

$$V(r) = \frac{-Ze^2}{r} e^{-s(r/\mu)} \tag{1.143}$$

where Z is the nuclear charge, the length μ is given by

$$\mu = 0.885 a_0 Z^{-1/2} \tag{1.144}$$

and s is an empirical constant. Such a potential is based on the Thomas-Fermi field for an atom [13]. Here s is taken as unity.

With this simple potential, the numerical problem reduces to solving the second-order ordinary differential equation (1.132) in exactly the same way as discussed in the previous subsection. At such high-incident electron energies, many partial waves may be required and one must be aware of the most economical way of computing a partial wave contribution, consistent with its being an accurate way. Since Born's approximation [13] is probably the fastest way of computing $\delta_l(k^2)$, then the question arises as to which $l \geq l_B$ can be calculated by this method. For $l \leq l_B$ we solve the differential equation. In practice for a fixed k^2 we compute δ_l from the differential equation for increasing l and about every fifth l calculate the Born value for δ_l also. When $|\delta_l - \delta_l^B|$ is less than some specified epsilon, then all subsequent higher partial waves can be computed using the Born approximation. In this way, we write

$$\cos \chi = \frac{1}{2k} \sum_{l=0}^{l_B} (2l + 1) \sin 2\delta_l + \frac{1}{2k} \sum_{l=l_B+1}^{L} (2l + 1) \sin 2\delta_l^B. \tag{1.145}$$

and similarly for $\sin \chi$, where δ_l^B is the Born value for the phase shift.

1.3.3. Differential Elastic Scattering of Molecular Beams [14]

A very simple model for the interaction of two molecules is to consider them as structureless particles which interact with each other according to the spherically symmetrical potential

$$V(r) = \varepsilon f\left(\frac{r}{\sigma}\right).$$

For such a model, the differential elastic scattering cross section can be computed using the formula of the preceding sections, but with some modifications in the numerical procedure if the potential is of the Lennard-Jones form

$$\bar{V}(x) = 4\left(\frac{1}{x^{12}} - \frac{1}{x^6}\right), \qquad x \equiv \frac{r}{\sigma}, \qquad \bar{V} = \frac{V}{\varepsilon}, \tag{1.146}$$

rather than the exponential potential. The most obvious modification to our method for solving the differential equation is that the singular behavior of

Eq. 1.146 is such that the integrations cannot be initiated at the origin. Bernstein assumed the existence of an infinite potential barrier at $x = x_s$. That is to say, he took a potential which had two parts: a hard core into which the projectile cannot penetrate and a long-range tail. Consequently, we have the physical boundary condition that

$$F_l(x_s) = 0$$

and since

$$\left[-k^2 + \frac{l(l+1)}{x_s^{\,2}} + U(x_s)\right]$$

will be finite, in general, the we must have

$$\left.\frac{d^2F_l}{dx^2}\right|_{x_s} = 0. \tag{1.147}$$

Since the differential equation is still homogeneous in $F_l(x)$, then its normalization is arbitrary as before and so we can choose $\left.\frac{dF_l}{dx}\right|_{x_s}$ to be any convenient number. The step-by-step numerical integration of the differential equation from x_s out into the domain of x where the potential can be neglected can be carried out as described later in Section 1.4. However, potentials which go to zero at infinity like $(1/x)^n$ can cause numerical inaccuracies. Numerical methods for handling such potentials have been developed and are discussed in Section 1.4.4.

A method for computing the phase shift, alternative to Eq. 1.139, is to compute the location of each zero of the numerical solution and compare it with the corresponding zero of the regular spherical Bessel function. The difference between the two sets of zeros settles down to a constant at those values of x for which $V(x)$ is negligible—this is the so-called asymptotic region. Bernstein found that for $x_s \leq 0.7$ his phase-shifts were independent of this choice of x_s. He also found that using the Runge-Kutta method for the numerical integration, the zeros and phases were independent of his choice of integration step, provided that step was less than 0.005.

The above two statements on x_s and H must be strongly emphasized. In using digital computers, there is usually less care and control exercised by the programmer than if a corresponding hand computation was carried out. In numerical computations it is essential to ensure that the results are indeed independent of those parameters which they are thought to be independent of—such as x_s, H, and $F_l'(x_s)$ in the problem above. It always has to be kept in mind that in the millions of operations performed by a computer in obtaining a result, round-off and truncation of formulas may cause a build up of error to such a level that there is no significance in the final answer. One of the most valuable aids in error detection, or fault-finding, in computer programs is to ensure that the results have the expected qualitative features. For example, as k^2 increases, the solution oscillates more rapidly and as l increases the solution becomes flatter at the origin. Further checking is possible if the numerical analysis used to solve a problem involves formulas which include terms that are normally neglected but can be included in check runs. Additional checking can be carried out by examining the numerical formulas for sensitive regions of subtractions, of possible singular or rapidly changing functions, and using double precision arithmetic in check runs and perhaps even including some double precision in the production runs of a computer code.

1.3.4. The Low-Temperature Properties of Gases

Helium persists as a monatomic gas down to temperatures so low that quantum effects become important. In calculating the viscosity, $\eta(T)$, and second virial coefficients, $B(T)$, of helium, Massey and Buckingham [15] and Buckingham et al. [16] have taken these effects into account. Both these quantities, η and B, can be expressed in terms of the energy dependence of the partial wave phase shift [15]. It is the computation of phase shifts which concerns us in this section. The potential $V(r)$ is now interpreted as the interaction between two helium atoms at a distance r apart. Massey and Buckingham chose the Slater form for the potential, that is

$$V(r) = (7.7 \times 10^{-10})e^{-(r/0.217)} - (1.47 \times 10^{-12})\frac{1}{r^6}\ \text{erg} \qquad (1.148)$$

Obviously at small distances the interaction between two helium atoms will be very complicated and Eq. 1.148 cannot represent the interaction. Once again the integration cannot be started from the origin. The procedure adopted by these authors was to compute the zero of

$$k^2 - \frac{l(l+1)}{r^2} + \frac{2m}{\hbar^2}V(r),$$

and call it r_0. At this value of the independent variable, $(d^2F_l/dr^2)r_0 = 0$. This determines one of the pair of integration constants; the other, $(dF_l/dr)_{r_0}$ say, must be chosen so that the function $F_l(r)$ vanishes at the origin. This was determined by trial and error. Once the slope is determined, then the equation was integrated out from r_0 step by step into the asymptotic region, where δ_l was determined.

The choice of the interatomic potential is of cardinal importance in the collision problem. The form in Eq. 1.148 was chosen because the r^{-6} has the form indicated theoretically for the leading term of the van der Waals energy, while the energy of the short-range repulsion is represented empirically by the exponential. Buckingham et al. [16] modified this form to

$$V(r) = be^{-ar} - cr^{-6} - dr^{-8} \tag{1.149}$$

which includes the second van der Waals term. The same type of potential was used by Buckingham, Davies, and Gilles [17] in calculating the viscosity cross section for hydrogen molecules.

These latter authors used a computer program to solve the second-order ordinary differential equation based on the Numerov method. The form of the potential used was

$$\frac{V}{\varepsilon} = -f(\sigma)$$

where

$$f(\sigma) = f_1 \frac{1}{\sigma^6}\left(1 + \frac{b}{\sigma^2}\right) - f_2 e^{-a(\sigma-1)},$$
$$f_2 = -1 + (1 + b)f_1 \quad \text{and} \quad f_1 = \frac{a}{a(1+b) - 6 - 8b}, \quad \sigma > 1 \tag{1.150}$$

For $\sigma < 1, f_1$ was multiplied by an exponential factor which removes the σ^{-6} and σ^{-8} terms. The numerical method used was the following:

Starting with a value of $\sigma = 0.5$, where the repulsive part of the potential is sufficiently large that the solution, F_l, is effectively zero, the differential equation was integrated outwards from the origin at an integration step of 0.01. If the short-range repulsive part of the potential had been approximated by a hard core, then the wave function must vanish identically at the hard-core radius. Since the solution, with positive k^2, is oscillatory, at each change of sign the position of the zeros, σ_k, where $F_l(\sigma_k) = 0$, was determined by interpolation. The outward integration of the differential equation was continued until the intermolecular potential fell below a preassigned level relative to the value of k^2. The phase shift was determined using

$$\frac{F_l(\sigma_m)}{k\sigma_m} = 0 = A_l j_l(k\sigma_m) - B_l n_l(k\sigma_m)$$

at the last pair of zeros of the solution. In general, these two estimates differed slightly from each other due to the residual effect of the long-range attractive tail, the r^{-6} term. A small correction was made using the WKB method, to be discussed in the next section. An alternative procedure would be to use the asymptotic method presented in Section 1.4.4.

1.3.5. The WKB Approximation

The WKB approximation, or Jeffrey's method, is discussed in nearly every textbook on quantum mechanics. Our reason for including it here is to draw attention to its recent use in collision problems. Let us begin by re-writing our basic radial equation in the form

$$\left[\frac{d^2}{dr^2} + w(r)\right]F(r) = 0, \tag{1.151}$$

and consider the scattering problem in which $w(r)$ tends to a positive constant at very large r that results in $F(r)$ being an oscillatory function. In order to take advantage of the oscillatory nature of $F(r)$, we shall write

$$F(r) = \zeta(r)^{-1/2} \sin \phi(r), \tag{1.152}$$

where ζ and ϕ are to be determined. We substitute Eq. 1.152 into 1.151 and invoke the fundamental theorem of algebra to set to zero the coefficients of $\cos \phi$ and $\sin \phi$, respectively, giving

$$\frac{d^2\phi}{dr^2} - \zeta^{-1}\frac{d\zeta}{dr}\frac{d\phi}{dr} = 0 \tag{1.153}$$

$$-\zeta^2 + w + \zeta^{1/2}\frac{d^2}{dr^2}\zeta^{-1/2} = 0. \tag{1.154}$$

A solution of Eq. 1.153 is

$$\zeta = \frac{d\phi}{dr}, \qquad \text{i.e.,} \qquad \phi(r) = \int_a^r \zeta(r)\, dr, \tag{1.155}$$

where a is an integration constant chosen to satisfy the boundary conditions. In other words, if we can determine $\zeta(r)$ in terms of the known function $w(r)$, using Eq. 1.154, then we can substitute the value into the integrand of Eq. 1.155 to obtain ϕ; consequently, we know $F(r)$! However, Eq. 1.154 is a nonlinear differential equation; one of the simplest methods of solving it is by iteration:

$$\zeta_n = \left[w + \zeta_{n-1}^{1/2}\frac{d^2}{dr^2}\zeta_{n-1}^{1/2}\right]^{1/2} \qquad \text{for} \quad n > 1, \tag{1.156}$$

while

$$\zeta_0 = w^{1/2}.$$

By transferring the oscillatory character of $F(r)$ into $\phi(r)$ we should expect that $\zeta(r)$ would be a very slowly varying function of r and so the iteration process might converge quite quickly.

Seaton and Peach [18] have computed the zero order approximation to ϕ for potentials of the form

$$w(r) = k^2 + \frac{\alpha - l(l+1)}{r^2} + \frac{2Z}{r} - u(r) \equiv w_0(r) - u(r), \tag{1.157}$$

where $u(r)$ vanishes asymptotically faster than the dipole potential r^{-2}. This is done by adding and subtracting a quantity ϕ_0 from the right-hand side of Eq. 1.155

$$\phi(r) = \phi_0(r) - \int_a^r [w_0^{1/2} - \zeta(x)]\, dx.$$

For very large r, we take $\phi(r)$ to be $\phi_0(r)$, which implies that the integral term is to vanish, hence

$$\phi(r) = \phi_0(r) + \int_r^\infty [w_0^{1/2} - \zeta(x)]\, dx, \tag{1.158}$$

where ϕ_0 will now be evaluated analytically and the integral can be evaluated numerically. If we define

$$X = ax^2 + bx - c, \qquad a = k^2, \qquad b = 2z, \qquad c = \alpha - l(l+1),$$

then according to Dwight [19]

$$\phi_0(r) = \int^r [k^2x^2 + 2zx + \alpha - l(l+1)]^{1/2} \frac{dx}{x} = X^{1/2} + \frac{b}{2}\int^r \frac{dx}{X^{1/2}} + c\int^r \frac{dx}{xX^{1/2}}\cdot$$

For the ranges, $a > 0$, and $c < 0$, we find

$$\phi_2(r) = X^{1/2} + \kappa^{-1}\log\left(1 + \kappa X^{1/2} + \frac{k^2 r}{Z}\right) + c^{1/2}\cos^{-1}\frac{[Xc(1 + c\kappa^2)^{-1}]^{1/2}}{rZ}$$
$$+ \text{(constant of integration)}, \tag{1.159}$$

where $\kappa = k/Z$. From the definition of ϕ_0, we see that it is in fact the phase of the Coulomb potential problem, consequently, we must have

$$\underset{r\to\infty}{\text{Lt}}\ \phi_0(r) = kr - \frac{l\pi}{2} + \frac{Z}{k}\log 2kr + \sigma_l, \tag{1.160}$$

where

$$\sigma_l = \arg\Gamma\left(l + 1 - i\frac{Z}{k}\right).$$

Consequently, when we take the asymptotic limit of both sides of Eq. 1.159; we determine the constant of integration to be

$$\sigma_l - \frac{l\pi}{2} - \kappa^{-1} - \kappa^{-1}\log\kappa - c^{1/2}\cos^{-1}\kappa\left(\frac{c}{1 + c\kappa^2}\right)^{1/2}.$$

Equation 1.159 is the zero-order WKB approximation to the phase as given by Seaton and Peach.

In practice it is often desirable to have an analytic solution to higher order than the zeroth. The first iteration of Eq. 1.156 gives

$$\begin{aligned}\zeta_1 &= \left(w_0 + w_0^{1/4}\frac{d^2}{dr^2}w_0^{-1/4}\right)^{1/2} \\ &= w_0^{1/2} + \tfrac{1}{2}w_0^{-1/4}\frac{d^2}{dr^2}w_0^{-1/4}\end{aligned} \tag{1.161}$$

if we neglect quadratic, and higher, terms in $w_0^{1/4}d^2(w_0^{-1/4})/dr^2$. The first-order phase, ϕ_1, is defined in terms of ζ_1, and integrating by parts gives

$$\phi_1 \equiv \int_a^r \zeta_1(x)\,dx = \int_a^r w_0^{1/2}\,dx + \left[\tfrac{1}{2}w_0^{-1/4}\frac{d}{dr}w_0^{-1/4}\right]_a^r - \frac{1}{2}\int_a^r \left(\frac{d}{dr}w_0^{-1/4}\right)^2 dr. \tag{1.162}$$

Problem 9. Perform the analytic integrations specified in Eq. 1.162 to obtain

$$\phi_1 = X^{1/2} + \kappa^{-1}\log(1 + X^{1/2}\kappa + r\kappa^2 Z) + \sigma_l - \kappa^{-1} - \kappa^{-1}\log\kappa$$
$$- \frac{l\pi}{2} + \Theta - \frac{X^{1/2}(3\kappa^2 c + 4) + \kappa r(3\kappa^2 c + 2)Z}{24(1 + \kappa^2 c)X^{1/2}(X^{1/2} + krz)} + \frac{5(rz - c)}{24X^{3/2}}$$

where Θ is defined in Burgess [20].

So far in this section we have been concerned with those solutions in Eq. 1.151 when $F(r)$ is oscillatory. Those solutions which decay exponentially are also of physical interest; indeed they take on a pivotal role in understanding many channel phenomena. Therefore, in addition to solutions of the form assumed in Eq. 1.152 we shall be interested in solutions

$$h(r) = \xi^{-1/2}e^{-\theta(r)}, \quad \text{for } r \text{ such that} \quad w < 0. \tag{1.163}$$

In order to compute $\theta(r)$ to the same approximation as $\phi(r)$ we substitute this equation back into Eq. 1.151 and relate to θ by

$$\xi = \frac{d\theta}{dr}, \quad \text{i.e.,} \quad \theta(r) = \int_b^r \xi(x)\,dx, \tag{1.164}$$

which results in the following equation for ξ, valid for any r,

$$\xi^2 + \xi^{1/2}\frac{d^2}{dr^2}\xi^{-1/2} + w = 0. \tag{1.165}$$

The constant of integration b is chosen such that

$$h(r) \underset{r\to\infty}{\sim} \exp\left(-|k|\, r + \frac{Z}{k}\log 2\,|k|\, r\right). \tag{1.166}$$

Problem 10. Determine θ_0 and θ_1 the zeroth and first-order approximants to θ.

1.4. SECOND-ORDER ORDINARY DIFFERENTIAL EQUATIONS

1.4.1. Numerov Method [21]

Before we can discuss this method we must derive some preliminary formulae. The first formula is Newton's formula for forward interpolation. The problem is to find a suitable polynomial for replacing any *given* function, $y(x)$, over a given interval, especially for interpolating a value of $y(x)$ near the end of tabulated values. Let $\phi(x)$ be a polynomial which we shall write as

$$\phi(x) = a_0 + a_1(x - x_n) + a_2(x - x_n)(x - x_{n-1}) + a_3(x - x_n)(x - x_{n-1}) \times (x - x_{n-2}) + \cdots + a_{n-1}(x - x_n)(x - x_{n-1})\cdots(x - x_2), \tag{1.167}$$

where x_i are the points at which $y(x_i)$ are tabulated. The coefficients a_j are chosen so as to make

$$\phi(x_i) = y_i, \qquad i = 1, 2, 3, \ldots, n,$$

hence

$$\phi(x_n) = a_0 = y_n;$$
$$\phi(x_{n-1}) = y_n + a_1(x_{n-1} - x_n) = y_{n-1}$$

or

$$a_1 = \frac{y_{n-1} - y_n}{x_{n-1} - x_n} \equiv \frac{\Delta_1 y_n}{h} \tag{1.168}$$

having introduced the notation $\Delta_1 y$ for first differences and h for the step size. At $x = x_{n-2}$, Eq. 1.167 yields

$$\phi(x_{n-2}) = y_n + \frac{\Delta_1 y_n}{h}(-2h) + a_2(-2h)(-h) = y_{n-2},$$

i.e.,

$$a_2 = \frac{y_{n-2} + 2\Delta_1 y_n - y_n}{2h^2}$$

$$= \frac{y_{n-2} - 2y_{n-1} + y_n}{2h^2} \equiv \frac{\Delta_2 y_n}{2h^2} \tag{1.169}$$

having defined the second difference operator Δ_2 by this last result. In general we shall have

$$a_n = \frac{1}{n!\,h^n}\Delta_n y_n, \tag{1.170}$$

see Table 1.

We should emphasize that we have used equally spaced steps throughout the range, that is, $x_i - x_{i-1} = h$. Substituting the values of a_j into Eq. 1.167 gives

$$\phi(x) = y_n + \frac{\Delta_1 y_n}{h}(x - x_n) + \frac{\Delta_2 y_n}{2h^2}(x - x_n)(x - x_{n-1})$$

$$+ \cdots + \frac{\Delta_{n-1} y_n}{(n-1)!\,h^{n-1}}(x - x_n)(x - x_{n-1})\cdots(x - x_2), \tag{1.171}$$

which is Newton's formula for backward interpolation written in terms of the independent variable x. This formula contains the values of y_i previous to y_n, that is, backward from the farthest point on the x-axis, and so can be used for interpolation at the end of a range. To simplify this formula we make a change of variable

$$u \equiv \frac{x - x_n}{h} \quad \text{or} \quad x = x_n + hu. \tag{1.172}$$

With this new independent variable we shall have

$$\frac{x - x_{n-1}}{h} = \frac{x - (x_n - h)}{h} = u + 1,$$

$$\frac{x - x_{n-2}}{h} = \frac{x - (x_n - 2h)}{h} = u + 2,$$

Table 1. Algebraic Formula for the First, Second, and Third Differences

x	y	$\Delta_1 y$	$\Delta_2 y$	$\Delta_3 y$
x_{n-4}	y_{n-4}			
		$y_{n-3} - y_{n-4}$		
x_{n-3}	y_{n-3}		$y_{n-2} - 2y_{n-3} + y_{n-4}$	
		$y_{n-2} - y_{n-3}$		$y_{n-1} - 3y_{n-2} + 3y_{n-3} - y_{n-4}$
x_{n-2}	y_{n-2}		$y_{n-1} - 2y_{n-2} + y_{n-3}$	
		$y_{n-1} - y_{n-2}$		$y_n - 3y_{n-1} + 3y_{n-2} - y_{n-3}$
x_{n-1}	y_{n-1}		$y_n - 2y_{n-1} + y_{n-2}$	
		$y_n - y_{n-1}$		$y_{n+1} - 3y_n + 3y_{n-1} - y_{n-2}$
x_n	y_n		$y_{n+1} - 2y_n + y_{n-1}$	
		$y_{n+1} - y_n$		
x_{n+1}	y_{n+1}			

and so on, which leads to Eq. 1.171 being written as

$$\phi(u) = y_n + \Delta_1 y_n \cdot u + \frac{\Delta_2 y_n}{2} \cdot u(u+1) + \frac{\Delta_3 y_n}{3!} \cdot u(u+1)(u+2)$$

$$+ \cdots + \frac{\Delta_{n-1} y_n}{(n-1)!} \cdot u(u+1)(u+2) \cdots (u+n-2) \quad (1.173)$$

We wish to compute the values of $y(x)$, for real positive x, which satisfy the second-order ordinary differential equation with first derivative absent,

$$\frac{d^2 y}{dx^2} = y'' = f(x, y), \quad (1.174)$$

together with two boundary conditions in order to specify the pair of integration constants. We take $y(x)''$ to be the function of x which we are approximating by the polynomial of Eq. 1.173:

$$y'' = y_n'' + u\,\Delta_1 y_n'' + \frac{u(u+1)}{2}\Delta_2 y_n'' + \frac{u(u+1)(u+2)}{6}\Delta_3 y_n'' + \cdots. \quad (1.175)$$

Since $x = x_n + hu$, where x_n and h are fixed, then

$$dx = h\,du \quad (1.176)$$

and upon integrating Eq. 1.174 with respect to x, we have

$$\int y''\,dx = h\int du\left\{y_n'' + u\,\Delta_1 y_n'' + \frac{u(u+1)}{2}\Delta_2 y_n'' + \cdots\right\}$$

which yields

$$y' = C_1 + h\left\{uy_n'' + \frac{u^2}{2}\Delta_1 y_n'' + \frac{1}{2}\left(\frac{u^2}{2} + \frac{u^3}{3}\right)\Delta_2 y_n'' + \cdots\right\},$$

where C_1 is the integration constant which is determined as follows: when $x = x_n$, that is, $u = 0$, then $y' \to y_n' = C_1$, consequently

$$y(x)' = y_n' + h\left\{uy_n'' + \frac{u^2}{2}\Delta_1 y_n'' + \left(\frac{u^2}{4} + \frac{u^3}{6}\right)\Delta_2 y_n'' + \cdots\right\}. \quad (1.177)$$

The method of replacing the derivative of a function by a polynomial and integrating that polynomial over an interval was used by J. C. Adams as early as 1855. We use the method a second time by integrating Eq. 1.177 with respect to x to obtain

$$y = huy_n' + h^2\left\{\frac{u^2}{2}y_n'' + \frac{u^3}{6}\Delta_1 y_n'' + \left(\frac{u^3}{12} + \frac{u^4}{24}\right)\Delta_2 y_n'' + \cdots\right\} + C_2.$$

The second integration constant C_2 is determined by noting that at $x = x_n$, where $u = 0$, we have $y \to y_n = C_2$. This gives

$$y(u) = y_n + huy_n' + h^2\left\{\frac{u^2}{2} y_n'' + \frac{u^3}{6}\Delta_1 y_n'' + \left(\frac{u^3}{12} + \frac{u^4}{24}\right)\Delta_2 y_n'' + \frac{1}{6}\left(\frac{u^3}{3} + \frac{u^4}{4} + \frac{u^5}{20}\right)\Delta_3 y_n'' + \frac{1}{24}\left(u^3 + \frac{11u^4}{12} + \frac{3u^5}{10} + \frac{u^6}{30}\right)\Delta_4 y_n'' + \cdots\right\} \tag{1.178}$$

If we set $u = -1, -2$ then we obtain the values of y at y_{n-1}, y_{n-2}, respectively, namely,

$$y_{n-1} = y_n - hy_n' + h^2\{\tfrac{1}{2}y_n'' - \tfrac{1}{6}\Delta_1 y_n'' - \tfrac{1}{24}\Delta_2 y_n'' - \tfrac{1}{45}\Delta_3 y_n'' - \tfrac{7}{480}\Delta_4 y_n'' + \cdots\}$$

$$y_{n-2} = y_n - 2hfy_n' + h^2\{2y_n'' - \tfrac{4}{3}\Delta_1 y_n'' + o\,\Delta_2 y_n'' - \tfrac{2}{45}\Delta_3 y_n'' - \tfrac{1}{30}\Delta_4 y_n'' + \cdots\}$$

Multiply the first equation by 2 and subtract the result from the second equation to get the truncated result

$$y_{n-2} - 2y_{n-1} \approx -y_n + h^2[y_n'' - \Delta_1 y_n'' + \tfrac{1}{12}\Delta_2 y_n'' - \tfrac{1}{240}\Delta_4 y_n'']$$

which can be rewritten as

$$y_n - 2y_{n-1} + y_{n-2} \approx h^2\,[y_{n-1}'' + \tfrac{1}{12}\Delta_2 y_n'' - \tfrac{1}{240}\Delta_4 y_n'']$$

If the last term on the right-hand side is neglected, then we have the result

$$y_n - 2y_{n-1} + y_{n-2} \approx h^2[y_{n-1}'' + \tfrac{1}{12}\Delta_2 y_n'']. \tag{1.179}$$

Second-order ordinary differential equations of the form

$$y'' = F(x) \cdot y + G(x) \tag{1.180}$$

where F, G are known functions, can be written as

$$y_j'' = F(x_j)y_j + G(x_j) \equiv F_i y_i + G_j$$

which is substituted into the right-hand side of Eq. 1.179, written with subscripts j rather than n, to obtain

$$y_j - 2y_{j-1} + y_{j-2} \approx h^2[F_{j-1}y_{j-1} + G_{j-1} + \tfrac{1}{12}\Delta_2\{F_j y_j + G_j\}],$$

which can be written as

$$y_j - 2y_{j-1} + y_{j-2} \approx h^2[F_{j-1}y_{j-1} + G_{j-1} + \tfrac{1}{12}(F_j y_j - 2F_{j-1}y_{j-1} + F_{j-2}y_{j-2}) + \tfrac{1}{12}\Delta_2 G_j].$$

Each subscript can be increased by unity to obtain

$$y_{j+1} - 2y_j + y_{j-1}$$
$$= h^2[F_j y_j + G_j + \tfrac{1}{12}(F_{j+1}y_{j+1} - 2F_j y_j + F_{j-1}y_{j-1}) + \tfrac{1}{12}\Delta_2 G_{j+1}].$$

Regrouping the terms of this equation gives the Numerov formula

$$[1 - \tfrac{1}{12}h^2F_{j+1}]y_{j+1} = 2[1 - \tfrac{1}{12}h^2F_j]y_j - [1 - \tfrac{1}{12}h^2F_{j-1}]y_{j-1}$$
$$+ h^2[F_j y_j + G_j + \tfrac{1}{12}\Delta_2 G_{j+1}] \quad (1.181)$$

For $j = 1$, we have a formula for determining y_2, provided we know the two previous ordinates y_0 and y_1. In the next section we discuss the task of computing these initial values.

A computer program for the numerical integration of Eq. 1.181 could involve the following steps:
Let $F1$ be the identifier for the current value of y, namely y_j
while $F2$ is the identifier for the previous value of y, namely y_{j-1}.

Let $F6$ be the identifier for the current value of $(1 - \tfrac{1}{12}h^2F_j)y_j$,
while $F7$ is the identifier for the previous value of $(1 - \tfrac{1}{12}h^2F_{j-1})y_{j-1}$.
Let $F8$ be the identifier for $h^2[F_j y_j + G_j + \tfrac{1}{12}\Delta_2 G_j]$.
that is,

$$F8 = 12(F1 - F6) + h^2[G_j + \tfrac{1}{12}\Delta_2 G_j]$$

We shall also assign an identifier, $F3$, to the factor multiplying y_{j+1}, and use this same identifier for the inverse of this factor, $[1 - \tfrac{1}{12}h^2F_{j+1}]^{-1}$.

In deriving Eq. 1.179 we neglected the term

$$-\frac{h^2}{240}\Delta_4 y_n \sim O(h^6).$$

As a consequence, it has long been assumed, quite erroneously, that the propagation error is $O(h^5)$. However, Sloan [22] proved that this is not so, but the propagation error is $O(h^4)$!

In Fig. 22 we have drawn a schematic of the integration scheme.

If the integration is begun at the origin, then for scattering problems $y(0) = 0$, so $F2 = 0$, while $F1$ can be computed using the methods of the

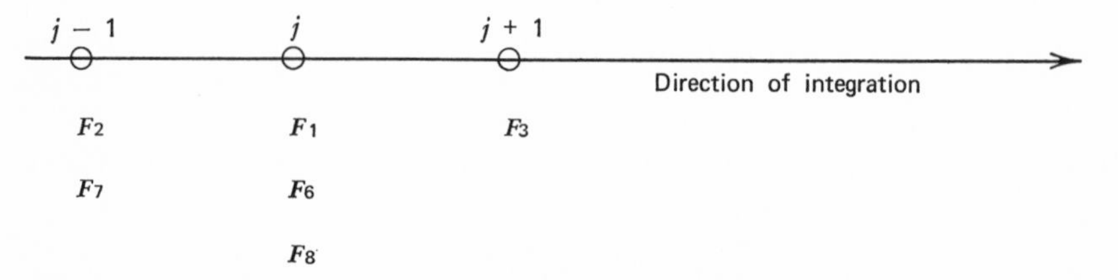

Figure 22 Schematic diagram of step-by-step integration.

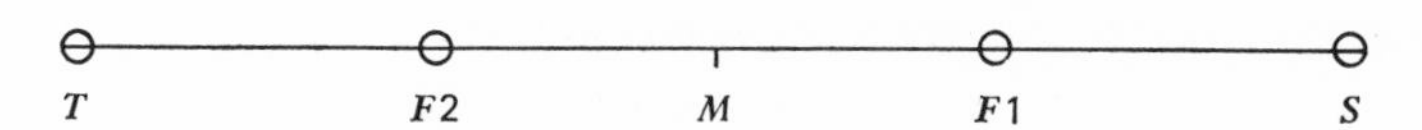

Figure 23 Half-intervals in the Numerov method.

next section. When $F2$ and $F1$ are known, then $F7$ and $F6$ can be computed. We can also compute $F8$ and $F3$; all the quantities marked on the figure are now known and we are in a position to compute y_{j+1}.

The computation can be carried out in the following sequence:

1. $F1 \rightarrow F2$, where the arrow denotes a storing operation.
2. Compute the right-hand side of Eq. 1.181 and store the result in $F1$, that is,

$$2*F6 - F7 + F8 \rightarrow F1.$$

3. $F6 \rightarrow F7$ and $F1 \rightarrow F6$. This makes $(j + 1)$ the current point.
4. Compute $F3^{-1} * F6 \rightarrow F1$, gives y_{j+1}.
5. Compute $F8$ at y_{j+1} and $F3$ at y_{j+2}

Repeat the whole procedure.

The Numerov method has two principal disadvantages: first the process of getting the method started, and second there is some difficulty in changing the mesh size. The advantage is that the formula (1.181) can be improved by including Δ_3 terms and so it is possible to keep track of the accuracy. A technique for dividing the mesh size by two would be the following:

Suppose we wish to halve the step from $F2$ outwards. To begin with, we integrate outward one more step h to the point S, and we must hold the point T in temporary storage, see Fig. 23. We can then use some central difference interpolation formula to get an accurate value for M, the midpoint of $(T, F2, F1, S)$. The value of M now becomes the new $F2$ for subsequent integrations at the new mesh size $h/2$. However, $F6$, $F7$, $F8$, and $F3$ are all h-dependent and so they must be recomputed at the new step size.

1.4.2. Starting the Solution

Consider the second-order differential equation for electron-hydrogen atom scattering in the Static Field Approximation, see Section 1.3.1, we have

$$\left[\frac{d^2}{dx^2} + k^2 - \frac{l(l+1)}{x^2} + 2\left(1 + \frac{1}{x}\right)e^{-2x}\right]F_l(kx) = 0 \qquad (1.182)$$

At small values of x, we assume that it is possible to expand the solution in an ascending power series in x

$$F_l(kx) = \sum_{n=0}^{\infty} a_n x^{n+\sigma} \qquad (1.183)$$

where σ is as yet unknown; σ determines the leading term in our series, and the expansion coefficients, a_n, are also unknown. Substitution of (1.183) into (1.182) gives

$$\sum_n a_n(n+\sigma)(n+\sigma-1)x^{n+\sigma-2} + k^2 \sum_n a_n x^{n+\sigma} - l(l+1) \sum_n a_n x^{n+\sigma-2}$$

$$+ 2(1+x^{-1})\left(1 - 2x + \frac{(-2x)^2}{2!} + \frac{(-2x)^3}{3!} + \cdots\right) \sum_n a_n x^{n+\sigma} = 0.$$

The coefficient of $x^{\sigma-2}$ leads to the indicial equation

$$a_0\{\sigma(\sigma-1) - l(l+1)\} = 0. \tag{1.184}$$

For $a_0 \neq 0$, we have

$$\sigma = l + 1 \qquad \text{or} \qquad -l,$$

that is,

$$F_l(kx) \underset{x\to 0}{\sim} x^{l+1} \qquad \text{or} \qquad x^{-l} \tag{1.185}$$

Since we want physically significant solutions, that is, we do not want infinite probabilities, we must take the former solution, since it is regular at the origin. It must be emphasized that this statement is equivalent to a boundary condition. We recall that the full radial function is defined by

$$R_l(r) \equiv \frac{F_l(kr)}{r} \underset{r\to 0}{\sim} x^l \qquad \text{or} \qquad x^{-l-1} \quad \text{according to (1.185).}$$

Hence the physical solution demands the solution which is regular at the origin

$$F_l(kx) \underset{x\to 0}{\sim} a_0 x^{l+1}, \qquad F_l(o) = 0. \tag{1.186}$$

We have said repeatedly that an ordinary second-order differential equation has two integration constants. If we were to equate all the coefficients of $x^{i+\sigma}$ to zero we should obtain recurrence relations for the a_i. These relations would show that all the coefficients $a_1, a_2, a_3, \ldots$ could be written in terms of a_0 and σ, which can be regarded as our pair of integration constants. Our boundary condition at the origin has fixed σ to be $l + 1$; consequently, only a_0 is to be determined. This parameter merely determines the normalization of $F_l(kx)$. We see from Eq. 1.182 that the differential equation is homogeneous in F, so once we have a solution we can multiply it by any x-independent complex constant and it will remain a solution. This means a_0 can remain arbitrary; for example, simply set $a_0 = 1$.

From Eq. 1.185 we see that the leading term in the first derivative is

$$\frac{dF_l}{dx} \underset{x\to 0}{\sim} a_0(l+1)x^l + O(x^{l+1}). \tag{1.187}$$

For $l = 0$, $F_l'(0) = a_0$, while for $l > 0$ we have $F_l'(0) = 0$. In other words, for $l \geq 1$ not only does the function vanish, but also its first derivative. In general, from (1.185) we have that for the lth partial wave, the function and its first l derivatives vanish; this accounts for the decreasing slopes of the Bessel functions at the origin for increasing order. However, to compute these higher derivatives more accurately we must look at the actual coefficients a_i. For example, to compute F_l'' we must evaluate the right-hand side of the differential equation as a limiting process

$$F_l'' \underset{x \to 0}{\sim} \{-k^2 + l(l+1)x^{-2} - 2x^{-1} + O(x^0)\}a_0 x^{l+1}$$

$$\underset{x \to 0}{\sim} a_0[l(l+1)x^{l-1} - 2x^l + O(x^{l+1})]$$

which equals $-2a_0$ for $l = 0$, $+2a_0$ for $l = 1$, and 0 for $l \geq 2$. We can summarize the above results into a table of values for starting the numerical integration of Eq. 1.182 from the origin $x = 0$. We see from Table 2 that if the integration is begun at $x = 0$ for $l > 1$, then F, F', and F'' are all zero and the solution remains zero, that is, we shall compute the trivial solution $F_l(x) = 0$ for all x. However, if we step a very small increment ε away from the origin, then F and its derivatives can be given the values computed from the first term of their series expansion. From the table we see that the solutions depend on our choice of the parameter a_0—the second of the integration constants. To obtain greater accuracy in the solution than that offered by the leading term at ε, one simply retains the next terms in the expansion.

From this section and its predecessor, (1.4.1), we have developed a numerical method for the step-by-step integration of the radial Schrödinger equation out from the origin into the asymptotic region where the potential $V(r)$ has a negligible effect. In this region, $x \geq x_A$ we know $F_l(kx_A)$ and its derivatives and so we can compute the phase shifts using Eq. 1.139.

Table 2. Starting Values of Functions and Derivatives in Terms of a_0

	$x = 0$			$x = \varepsilon$
l	0	1	>1	≥1
F_l	0	0	0	$a_0\varepsilon^{l+1}$
F_l'	a_0	0	0	$a_0(l+1)\varepsilon^l$
F_l''	$-2a_0$	$2a_0$	0	$a_0 l(l+1)\varepsilon^{l-1}$

1.4.3. Matching Procedure for Bound States

Equation 1.182 can be written as

$$\frac{d^2F_l}{dx^2} + \left[E - \frac{l(l+1)}{x^2} + V(x)\right]F_l = 0, \tag{1.188}$$

where the energy E for collision problems is known and is positive. In Section 1.2.1 we considered this equation for attractive $V(x)$ and E negative and correlated the solutions F_l which vanished asymptotically with the bound states of the system. We compare this vanishing at infinity for bound states with the oscillatory character for scattering states and conclude that while the latter can still be a linear superposition of the two independent solutions (sine and cosine) the former can only have the solution which vanishes as $e^{-E^{1/2}x}$. In other words, the coefficient of $e^{E^{1/2}x}$ must vanish! So we see that for bound state problems we have two-point boundary conditions: (1) solution must be regular at the origin, (2) solution must vanish at infinity.

The method for starting the solution, discussed in the preceding section, involved complete freedom of choice in the slope of $F(x)$ at the origin. This is possible because when we get to infinity all we need is the ratio of the coefficients of sine and cosine, not their actual magnitudes. For bound state problems, if we choose the slope at the origin to be any value, then most certainly some of the solution $e^{E^{1/2}x}$ would be part of the numerical solution! Alternatively, since we know

$$F_l(E, x) \underset{x\to\infty}{\sim} Ce^{-E^{1/2}x}, \quad C \text{ unknown amplitude,}$$

then

$$\left[F_l^{-1}\frac{dF_l}{dx}\right]_{x\geq x_A} = -(E)^{1/2}. \tag{1.189}$$

From this result we see that in the asymptotic domain, the logarithmic derivative is known. That is, one of the pair of integration constants is known at $x \geq x_A$; however, the other is unknown. If we were to choose any value for $F_l(x_A)$, for example $e^{-E^{1/2}x_A}$, then integrate the differential equation backwards step by step to the origin, we would miss the origin, that is, $F_l(0) \neq 0$, thereby violating the boundary condition at the origin.

We are interested in the bound state problem for the following reason: in the many-channel problem we shall encounter many equations like (1.182) all coupled together, the different functions will be designated $F_\nu(x)$, $\nu = 1, 2, 3, \ldots, N$ with different, in general, energies k_ν^2, some of which will be positive and some negative. It is the latter which will be handled like the bound states of this section. The method of handling the bound states is based on the simple fact that the method must guarantee $F_l(0) = 0$ and Eq. 1.189. Consequently, starting at the origin, with the boundary condition there satisfied, one chooses anything for the slope and integrates outward from the origin to some point, x_0, where $V(x)$ is not too small compared with E. The outward integration is stopped at x_0. In those problems where E is known the step-by-step method is initiated at $x = x_A$, where Eq. 1.189 is imposed and any choice of $F_l(x_A)$ is assumed, and the integration is carried inward to x_0, where it is stopped. Except for a fluke, the outward value of the solution, $\mathscr{F}_l(x_0)$, will not equal the inward value of the solution, labelled $\mathscr{G}_l(x_0)$. However, by imposing the continuity of the solutions at x_0, henceforth to be called the matching point, we have

$$u\mathscr{F}_l(x_0) = \mathscr{G}_l(x_0), \tag{1.190}$$

where u, the matching parameter, can be determined by this equation. In other words, when E is known and negative, the full continuous solution with the correct boundary conditions are

$$\frac{\mathscr{G}_l(x_0)}{\mathscr{F}_l(x_0)} \cdot \mathscr{F}_l(x) \qquad \text{for} \qquad x < x_0.$$

and

$$\mathscr{G}_l(x) \qquad \text{for} \qquad x > x_0.$$

It is well known that for negative E there is not a continuous distribution of solutions for any E as there is for the scattering states, $E > 0$. Solutions exist only at certain discrete values of negative E; unless we use such E in the above discussion then we shall not get a smooth solution. Indeed, the numerical solution of the bound state problem involves determining the eigenvalue E. In Chapter 2 when we consider the many-channel problem and its connection to resonances we see that we can get an approximation to E_{res} by solving the truncated set of coupled radial equations with $E_\nu < 0$.

The method which we describe for determining *both* the eigenvalue and the matched solution was devised by Ridley [23]. The method enables an estimate of ΔE required to produce a match to be estimated from the degree of the mismatch of the solutions at $x = x_0$. Consider Eq. 1.188 and let $F + \Delta F$ be a solution when E is replaced by $E + \Delta E$, so that, to first order in small quantities:

$$\left[\frac{d^2}{dx^2} + V(x) + E - \frac{l(l+1)}{x^2}\right] \Delta F_l = -F_l \Delta E, \tag{1.191}$$

Equations 1.188 and 1.191 are multiplied by $-\Delta F_l$ and F_l respectively and added together to give

$$F_l \frac{d^2 \Delta F_l}{dx^2} - \Delta F_l \frac{d^2 F_l}{dx^2} = -F_l^{\,2} \Delta E. \tag{1.192}$$

The left-hand side can be written in the form

$$\frac{d}{dx}\left[F_l \frac{d}{dx} \Delta F_l - \Delta F_l \frac{dF_l}{dx}\right].$$

Since ΔF is the difference between two functions at the same value of r, then

$$\frac{d}{dr} \Delta F \equiv \Delta(F_l').$$

When these results are substituted into the left-hand side of Eq. 1.192 and the result integrated over the range a to b, we obtain

$$\int_a^b d[F_l \Delta(F_l') - \Delta F_l F_l'] = -\int_a^b dx F_l^2 \Delta E$$

which yields ΔE in terms of the solutions, namely,

$$\frac{[F_l \Delta(F_l') - \Delta F_l F_l']_a^b}{\int_a^b dx F_l^2(x)} = -\Delta E \tag{1.193}$$

Equation 1.193 is evaluated for the integration from the origin, $a = 0$ and $b = x_0$, where

$$\mathscr{F}_l(a) = 0,$$

and there is no error in the solution at the origin, hence

$$\Delta F_l = 0.$$

Consequently, we have

$$-\Delta E = \frac{[F_l \Delta(F_l') - F_l' \Delta F_l]_{x_0(\text{outw})}}{\int_0^{x_0} dx \mathscr{F}_l(x)^2} \tag{1.194}$$

Equation 1.193 is now evaluated for the integration from the asymptotic domain, $a = \infty$ to the match point, $b = x_0$. Since the solution vanishes asymptotically, then

$$\mathscr{G}_l(a) = 0$$

while there is no error at infinity, hence

$$\Delta F_l = 0.$$

These results lead to a second formula for the error in the eigenvalue, namely

$$-\Delta E = \frac{[F_l \Delta F_l' - F_l' \Delta F_l]_{x_0(\text{inw})}}{-\int_{x_0}^{\infty} dx \mathscr{G}_l^2(x)} \tag{1.195}$$

If we use the identity

$$F\,\Delta F' - F'\,\Delta F = F^2\,\Delta\left(\frac{F'}{F}\right),$$

then Eqs. 1.194 and 1.195 can be rewritten as

$$-\Delta E\int_0^{x_0}\frac{dx\mathscr{F}_l^{\,2}(x)}{\mathscr{F}_l^{\,2}(x_0)} = \Delta\left(\frac{F'}{F}\right)_{\text{outw}} = \Delta\left(\frac{\mathscr{F}'}{\mathscr{F}}\right)$$

$$\Delta E\int_{x_0}^{\infty}\frac{dx\mathscr{G}_l^{\,2}(x)}{\mathscr{G}_l^{\,2}(x_0)} = \left(\frac{F'}{F}\right)_{\text{inw}} = \Delta\left(\frac{\mathscr{G}'}{\mathscr{G}}\right) \tag{1.196}$$

We assume that inward and outward integrations have been carried out for some trial value $E^{(0)}$ and gives values, labelled $(\mathscr{F}'/\mathscr{F})^0$ and $(\mathscr{G}'/\mathscr{G})^0$, that do not match at x_0. We want to make such a change ΔE of E such that

$$\left(\frac{\mathscr{G}'}{\mathscr{G}}\right)^0 + \Delta\left(\frac{\mathscr{G}'}{\mathscr{G}}\right) = \left(\frac{\mathscr{F}'}{\mathscr{F}}\right)^0 + \Delta\left(\frac{\mathscr{F}'}{\mathscr{F}}\right). \tag{1.197}$$

When Eq. 1.196 is substituted into (1.197) we obtain

$$\left(\frac{\mathscr{G}'}{\mathscr{G}}\right)^0 - \left(\frac{\mathscr{F}'}{\mathscr{F}}\right)^0 = \Delta E\left[-\int_0^{x_0}\frac{dx\mathscr{F}_l^{\,2}(x)}{\mathscr{F}_l^{\,2}(x_0)} - \int_{x_0}^{\infty}\frac{dx\mathscr{G}_l^{\,2}(x)}{\mathscr{G}_l^{\,2}(x_0)}\right], \tag{1.198}$$

which determines ΔE in terms of the computed values of the eigenfunction. The process will have to be repeated iteratively.

1.4.4. Asymptotic Expansion

In our discussion of the numerical procedures to be used to solve second-order ordinary differential equations we have assumed that the potentials $V(x)$ tend to zero exponentially and that at x_A, the value of the independent variable beyond which $V(x)$ can be neglected, the numerical solution can be fitted to Bessel functions. However, we have seen in Sections 1.3.3 and 1.3.4 that multipole potentials, potentials which vanish at infinity as x^{-s}, $s \geq 1$, are especially important. The value of x_A would be very great indeed if we just adopted the step-by-step procedures discussed above. This would have the following undesirable consequences: (1) greater amount of computer time required to compute each partial wave phase shift, (2) the accumulative effects of truncation in the numerical formulas and round-off in the arithmetic operations may build up to such a level that all significance is lost in the final result. This would mean that a code which had been thoroughly checked out for exponential potentials might well give inaccurate results with the long-range potentials. Indeed, we remarked at the end of Section 1.3.3 that the WKB method had been used to make a correction to δ_l after stopping the integration at some x_A where $V(x_A)$, although small, was not negligible. We now describe a numerical procedure due to Burke and Schey [24] for keeping x_A at about the same value for all potentials, except the Coulomb potential. The technique was extended to include the Coulomb potential by Burke, McVicar, and Smith [25].

In typical problems $U(x)$ has two types of terms, one of which is called short-range forces, usually involving an exponential factor, and long-range forces, involving inverse powers of x, that is,

$$U(x) = Ae^{-\mu x} \sum_n b_n x^n + \sum_{\lambda=1}^{\Lambda} a_\lambda x^{-\lambda}. \tag{1.199}$$

At very low energies the projectile will come under the influence of the long-range forces, and if they are repulsive the projectile may never penetrate close enough to those values of x where the short-range forces can become effective. As the energy of the projectile increases it will penetrate deeper and deeper into the target and it will be the short-range potential which will determine the principal characteristics of the scattering. The centrifugal force, $l(l+1)/x^2$, is an example of a long-range repulsive potential.

We know that if we integrate the radial equation out to $x \geq x_A$, where the short-range forces can be neglected, we shall then be solving the equation

$$\frac{d^2F_l}{dx^2} + k^2 - \left[\frac{l(l+1)}{x^2} - \sum_{\lambda=0}^{\Lambda} \frac{a_\lambda}{x^{\lambda+1}}\right] F_l(x) = 0. \tag{1.200}$$

From what we have said earlier, it is obvious that this equation will have to be integrated out to x far greater than x_A before F_l can be fitted to the Bessel functions. Indeed we recall that if $\lambda = 0$ was allowed, we should never be able to make such a fit, we should have to use the Coulomb functions of Section 1.1.5. If $\lambda = 1$ only appeared, then we could make a fit to spherical Bessel functions of noninteger order ν given by

$$\nu(\nu + 1) = l(l+1) + a_1. \tag{1.201}$$

The method of Burke and Schey is based on expanding the solution of the differential equation in a descending power series in x, in a somewhat analogous way to the method of starting the solution, but with a difference—the oscillatory nature of F_l is demonstrated explicitly

$$F_l(kx) = \sin\theta \sum_{p=0}^{\infty} \alpha_p x^{-p} + \cos\theta \sum_{p=0}^{\infty} \beta_p x^{-p}, \tag{1.202}$$

where the x-dependence of the phase angle θ can be taken to be

$$\theta = kx + \eta \log 2kx, \tag{1.203}$$

see Eq. 1.63, where we defined $\eta = Z_1Z_2/k$. For electrons scattered by positive ions we have $Z_1 = -1$ and $Z_2 = (Z - N)$, where Z is the nuclear charge and

N the number of bound electrons on the target ion. Hence $\eta = -(Z - N)/k$, so if we change our definition of η to be $(Z - N)/k$, then we must change the sign of the log term in Eq. 1.203, and that is what we have done in Eq. (1.63). Burke and Schey merely retained the first term in Eq. 1.203, as they were considering electrons scattering by neutral hydrogen atoms. The analysis here is for the more general case of scattering by charged particles when there is a residual Coulomb interaction, and it will include the neutral case as a special case by setting $\eta = 0$ in the final formulas.

A few further remarks on the choice of θ can be made here. Equations 1.202 and 1.203 are one way of expressing the oscillatory character of F_l; in general we can write

$$F_l(kx) = a_l(x)\sin\theta_l(x) + b_l(x)\cos\theta_l(x) \tag{1.204}$$

where as yet the amplitude and phase-functions are unspecified. The choice of Eq. 1.203 implies a distant x and "throws" the remaining unknown characteristics of the solution onto the coefficients α_p and β_p. If an improved functional form can be given for $\theta_l(x)$, as we did in Section 1.3.5 in our discussion of the WKB approximation, then we are leaving less of the characteristics of the solution to the amplitudes $a_l(x)$ and $b_l(x)$.

In order to determine the parameters α_p and β_p, the assumed asymptotic form (1.202) is substituted into Eq. 1.200. We can readily show that

$$\begin{aligned}\frac{d^2F_l}{dx^2} &= \sum_p \bigg\{p(p+1)x^{-p-2}[\alpha_p\sin\theta + \beta_p\cos\theta] + 2px^{-p-1}\left(k + \frac{\eta}{x}\right)\\ &\quad \times [-\alpha_p\cos\theta + \beta_p\sin\theta] + \eta x^{-p-2}[-\alpha_p\cos\theta + \beta_p\sin\theta]\\ &\quad + x^{-p}\left(k + \frac{\eta}{x}\right)^2[-\alpha_p\sin\theta - \beta_p\cos\theta]\bigg\}\\ &= \left\{-k^2 + \frac{l(l+1)}{x^2} + \sum_{\lambda=0}^{\Lambda}\frac{a_\lambda}{x^{\lambda+1}}\right\}F_l \end{aligned}\tag{1.205}$$

In order to determine α_p and β_p we equate powers of $\sin\theta x^{-p}$ and $\cos\theta x^{-p}$ on both sides of Eq. 1.205. The former yields

$$\begin{aligned}&\alpha_{p-2}(p-2)(p-1) - k^2\alpha_p - 2\eta k\alpha_{p-1} - \eta^2\alpha_{p-2} + k^2\alpha_p - l(l+1)\alpha_{p-2}\\ &\quad - a_0\alpha_{p-1} + 2(p-1)k\beta_{p-1} + 2(p-2)\eta\beta_{p-2} + \eta\beta_{p-2} = \sum_{\lambda=1}^{p-1} a_\lambda\alpha_{p-\lambda-1},\end{aligned}$$

where $a_0 = -2\eta k$, while the latter yields

$$\begin{aligned}&\beta_{p-2}(p-2)(p-1) - 2(p-1)k\alpha_{p-1} - 2(p-2)\eta\alpha_{p-2} - \eta\alpha_{p-2} - k^2\beta_p\\ &\quad - 2k\eta\beta_{p-1} - \eta^2\beta_{p-2} + k^2\beta_p - l(l+1)\beta_{p-2} - a_0\beta_{p-1} = \sum_{\lambda=1}^{p-1} a_\lambda\beta_{p-\lambda-1}.\end{aligned}$$

The pair of equation can be simplified to the forms

$$\alpha_{p-2}[(p-2)(p-1)-\eta^2-l(l+1)]+2(p-1)k\beta_{p-1}+(2p-3)\eta\beta_{p-2} = \sum_{\lambda=1}^{p-1} a_\lambda \alpha_{p-\lambda-1}, \quad (1.206)$$

and

$$\beta_{p-2}[(p-2)(p-1)-\eta^2-l(l+1)]-2(p-1)k\alpha_{p-1}-(2p-3)\eta\alpha_{p-2} = \sum_{\lambda=1}^{p-1} a_\lambda \beta_{p-\lambda-1}.$$

This is a pair of recurrence relations which are coupled together and allow the computation of all α_p and β_p, provided α_0 and β_0 are known. In other words, as we should expect, the descending-power series solution to the second-order differential equation in the asymptotic domain involves two constants of integration α_0 and β_0, and their ratio yields the phase shift. Let us take a closer look at how Eqs. 1.206 work: For $p = 2$ in the first relation, the first term involves α_0, the second term involves β_1, the third term β_0, and the right-hand side α_0; this allows the calculation of β_1, provided α_0 and β_0 are known. In the second relation, when $p = 2$, the first term involves β_0, the second α_1, the third α_0, and the right-hand side β_0; hence we can calculate α_1 knowing β_0 and α_0. We can readily see that with increasing p we have to calculate the α_p and β_p alternately. As the coefficients α_p and β_p are calculated, the contribution of that term in the descending series (1.202) is calculated and the recurrence relations are used until the contribution is less than some epsilon. Let us emphasize that this computation is carried out at some $x > x_A$, but as close to x_A as possible; call this point x_B. In practice, x_B is chosen such that less than ten terms are required in the descending power series to get a value of F_l within the specified epsilon.

We have yet to discuss how α_0 and β_0 are determined. Since α_0 and β_0 are two linearly independent parameters, then if we make a pair of choices for them, ($\alpha_0 = 1$, $\beta_0 = 0$) and ($\alpha_0 = 0$, $\beta_0 = 1$) say, we can construct any solution of the differential equation by a linear superposition of our pair of linearly independent choices. That is to say, with these choices we can now calculate $F_l(x_B)$ and $F_l(x_B + H)$, using (1.202), provided less than ten terms are required, where H is the step size for numerical integration. If more than ten terms are required, *increase* x_B and try again; a cut-off should be included in any computer program to prevent this process from continuing indefinitely. Indeed, if the cut-off is reached, then the asymptotic expansion method as described above has failed and an alternative procedure is required. Knowing $F_l(x_B)$ and $F_l(x_B + H)$, the Numerov method can be used to integrate the differential equation, *twice* inwardly to x_0. For collision problems, when the F are oscillatory we have shown that it is necessary to integrate outwards, once only, from the origin to the asymptotic domain. For a single open channel scattering problem $x_A \equiv x_B = x_0$.

To conclude: it is now clear that if we compute two linearly independent solutions of the differential equation in the asymptotic region from the descending series, labelled

$$\mathscr{G}_l^{(1)}(x_B) \quad \text{for} \quad \alpha_0 = 1, \quad \beta_0 = 0,$$

and

$$\mathscr{G}_l^{(2)}(x_B) \quad \text{for} \quad \alpha_0 = 0, \quad \beta_0 = 1,$$

then solution $\mathscr{F}_l(x)$ obtained by the numerical integration of the differential equation out from the origin to x_B must be a linear combination.

$$\mathscr{F}_l(x_B) = k^{-1/2}\{A_l \mathscr{G}_l^{(1)}(x_B) + B_l \mathscr{G}_l^{(2)}(x_B)\}, \tag{1.207}$$

where $A_l(k)$ and $B_l(k)$ are the only unknowns and their ratio determines $\tan \delta_l(k)$. Notice that a factor $k^{-1/2}$, common to both A and B, has been pulled out. If we take A_l to be unity, then

$$\tan \delta_l(k) = B_l(k) = \frac{k^{1/2}\mathscr{F}_l(x_B) - \mathscr{G}_l^{(1)}(x_B)}{\mathscr{G}_l^{(2)}(x_B)}. \tag{1.208}$$

For the single-channel problem, the elastic scattering threshold, $k^2 = 0$, is a particularly interesting energy region. At this threshold, it is quite clear that the oscillatory function of Eq. 1.202 cannot represent the asymptotic solution to the equation

$$\frac{d^2F_l}{dx^2} - \left[\frac{l(l+1)}{x^2} + \sum_{\lambda=0}^{\Lambda} \frac{a_\lambda}{x^{\lambda+1}}\right] F_l(x) = 0. \tag{1.209}$$

In the absence of the Coulomb potential, $a_0 \equiv 0$, the characteristics of the solution in the asymptotic domain can be found by trying a descending series expansion

$$F_l(x) = \sum_{p=0}^{\infty} \delta_p x^{\sigma-p}, \tag{1.210}$$

which when substituted into Eq. 1.209 yields

$$\sum_p \delta_p(\sigma - p)(\sigma - p - 1)x^{\sigma-p-2} - l(l+1)\sum_p \delta_p x^{\sigma-p-2}$$
$$- \sum_{\lambda,p} a_\lambda \, \delta_p x^{\sigma-p-\lambda-1} = 0.$$

When the coefficient of $x^{\sigma-2}$ is set equal to zero, we obtain for $\delta_0 \neq 0$,

$$\sigma(\sigma - 1) - \{l(l+1) + a_1\} = 0,$$

that is

$$\sigma = \nu + 1 \quad \text{or} \quad -\nu, \quad \text{where } \nu(\nu+1) \equiv l(l+1) + a_1.$$

The recurrence relations are obtained by setting the coefficient of $x^{\sigma-p-2}$ to zero, namely,

$$\delta_p[(\sigma - p)(\sigma - p - 1) - l(l+1)] - \sum_{\lambda=1}^{\Lambda} a_\lambda \, \delta_{p-\lambda+1} = 0, \quad p \geq 1.$$

From the indicial equation we saw that δ_0 was arbitrary and the recurrence relations give δ_p^+, when $\sigma = \nu + 1$, and δ_p^- when $\sigma = -\nu$.

In other words, asymptotically we have the linear superposition

$$F_l(x) \sim \sum_{p=0}^{\infty} \delta_p^+ x^{\nu+1-p} + \sum_{p=0}^{\infty} \delta_p^- x^{-\nu-p}. \tag{1.211}$$

As for positive energies, we must extract two linearly independent solutions in the asymptotic domain by setting ($\delta_0^+ = 1$, $\delta_0^- = 0$) and ($\delta_0^+ = 0$, $\delta_0^- = 1$), respectively. As before, we can label these solutions $\mathscr{G}_l^{(1)}$ and $\mathscr{G}_l^{(2)}$ and fit them to the outward solution $\mathscr{F}_l$, using Eq. 1.208.

1.4.5. Runge-Kutta Method [26]

We have remarked earlier that one of the disadvantages of the Numerov method is starting the solution. The great advantage of the Runge-Kutta method is that it is self-starting; given the function and derivative of a second-order ordinary differential equation at any x_i, the formulas allow the computation of $F_l(x_i + H)$ and $F_l'(x_i + H)$. This makes it possible to change the step size at any point in the integration. This is important because it saves computer time; if the potentials are varying slowly, then the solutions will also vary slowly and so a reasonably large step length can be used; when the potentials vary rapidly then smaller steps H are required. As a rough rule of thumb about twenty steps should be made throughout any loop of an oscillatory function. This implies that as k^2 increases and the frequency of the oscillations increase H must be decreased; similarly H must be decreased when the residual Coulomb charge in collision processes increases. The principal disadvantage of the Runge-Kutta method is that the error incurred in truncating the formulas, which are usually correct to fourth-order in H, is difficult to assess. Another disadvantage is that the derivatives must be computed several times within each step H.

Consider the first-order ordinary differential equation

$$\frac{dy}{dx} = f(x, y) \tag{1.212}$$

together with the boundary condition $y = y_0$ at $x = x_0$. It is assumed that a solution exists which can be developed in a Taylor series about x_0, namely,

$$\delta y \equiv y - y_0 = h\left[\frac{dy}{dx}\right]_{x_0} + \frac{h^2}{2!}\left[\frac{d^2y}{dx^2}\right]_{x_0} + \frac{h^3}{3!}\left[\frac{d^3y}{dx^3}\right]_{x_0} + \cdots. \tag{1.213}$$

The values of the derivatives of y can be expressed in terms of the known function $f(x, y)$, since

$$\left[\frac{dy}{dx}\right]_{x_0} = f(x_0, y_0),$$

and

$$\frac{d}{dx}\left(\frac{dy}{dx}\right) = \frac{d}{dx} f(x, y) = \frac{\partial f}{\partial x}\frac{\delta x}{dx} + \frac{\partial f}{\partial y}\frac{\delta y}{dx}$$

$$= \frac{\partial f}{\partial x} + f\frac{\partial f}{\partial y} \equiv Df. \tag{1.214}$$

Similarly, the third derivative can be written in the form

$$\frac{d}{dx}\left(\frac{d^2y}{dx^2}\right) = \frac{d}{dx}\left(\frac{\partial f}{\partial x} + f\frac{\partial f}{\partial y}\right)$$

$$= \frac{1}{dx}\left[\frac{\partial}{\partial x}\left(\frac{\partial f}{\partial x}\right)\delta x + \frac{\partial}{\partial y}\left(\frac{\partial f}{\partial x}\right)\delta y + \frac{\partial}{\partial x}\left(f\frac{\partial f}{\partial y}\right)\delta x + \frac{\partial}{\partial y}\left(f\frac{\partial f}{\partial y}\right)\delta y\right]$$

$$= \frac{\partial^2 f}{\partial x^2} + 2f\frac{\partial^2 f}{\partial x\,\partial y} + f^2\frac{\partial^2 f}{\partial y^2} + \frac{\partial f}{\partial y}\left(\frac{\partial f}{\partial x} + f\frac{\partial f}{\partial y}\right)$$

$$\frac{d^3y}{dx^3} \equiv (D^2 + f_y D)f, \tag{1.215}$$

which defines both D^2 and f_y. When this process is continued for the fourth derivative of y we can rewrite the Taylor series wholly in terms of f and its partial derivatives, that is,

$$\delta y = hf + \frac{h^2}{2!}Df + \frac{h^3}{3!}(D^2 f + f_y Df)$$

$$+ \frac{h^4}{4!}(D^3 f + f_y D^2 f + f_y{}^2 Df + 3Df Df_y) + \cdots. \tag{1.216}$$

Let us define the number

$$k_1 \equiv hf(x_0 y_0), \tag{1.217}$$

which from our defining equation we see gives a zero-order estimate of δy when δx is set equal to h,

$$\delta y \approx k_1.$$

At some point within the step h, at $(x_0 + \alpha h)$ say, where $\alpha < 1$, we assume that the ordinate at that point is $(y_0 + \beta k_1)$. From the defining equation we can evaluate the derivative at this point

$$\left[\frac{dy}{dx}\right]_{x_0+\alpha h} = f(x_0 + \alpha h, y_0 + \beta k_1),$$

and by analogy with Eq. 1.217 we define an alternative estimate of δy

$$k_2 \equiv hf(x_0 + \alpha h, y_0 + \beta k_1). \tag{1.218}$$

We now have two estimates of the increment in ordinate, δy, over a mesh point. A third estimate can be defined by

$$\delta y \approx R_1 k_1 + R_2 k_2,$$

where R_1 and R_2 are, as yet unknown, weighting factors.

At a second point within the mesh, $(x_0 + \alpha_1 h)$ say, the ordinate can be taken to be $(y_0 + \beta_1 k_1 + \gamma_1 k_2)$, while the derivative at this point is given by Eq. 1.212 to be

$$\left[\frac{dy}{dx}\right]_{x+\alpha_1 h} = f(x_0 + \alpha_1 h, y_0 + \beta_1 k_1 + \gamma_1 k_2),$$

which leads to the definition

$$k_3 \equiv hf(x_0 + \alpha_1 h, y_0 + \beta_1 k_1 + \gamma_1 k_2), \tag{1.219}$$

and yet a further estimate of the increment in the ordinate

$$\delta y = R_1 k_1 + R_2 k_2 + R_3 k_3.$$

This procedure can be carried out yet again for the point $(x_0 + \alpha_2 h)$ with

$$k_4 \equiv hf(x_0 + \alpha_2 h, y_0 + \beta_2 k_1 + \gamma_2 k_2 + \delta_2 k_3), \tag{1.220}$$

and a final estimate of the increment being

$$\delta y = \sum_{i=1}^{4} R_i k_i. \tag{1.221}$$

To compute δy from Eq. 1.221 we require the four weights, R_i, and the k_i can be evaluated if we knew the nine constants, α, β, α_1, β_1, γ_1, α_2, β_2, γ_2, and δ_2.

We note that we have two formulas for δy, namely Eqs. 1.216 and 1.221. We shall evaluate the thirteen unknowns by ensuring that this pair of estimates of δy agree with one another up to and including the term in h^4. To do this the expressions for k_2, k_3, and k_4 are expanded in powers of h, using the Taylor expansion in two variables, namely,

$$f(x + p, y + q) = f(x, y) + Df(x, y) + \frac{1}{2!} D^2 f(x, y) + \frac{1}{3!} D^3 f(x, y) + \cdots,$$

where

$$Df \equiv \left(p \frac{\partial}{\partial x} + q \frac{\partial}{\partial y}\right) f,$$

and

$$D^r f \equiv \left(p \frac{\partial}{\partial x} + q \frac{\partial}{\partial y}\right)^r f.$$

When these results are used in Eq. 1.218 we have

$$\begin{aligned} k_2 &= h\left[f(x_0, y_0) + \left(\alpha h \frac{\partial}{\partial x} + \beta k_1 \frac{\partial}{\partial y}\right) f + \frac{1}{2!}\left(\alpha h \frac{\partial}{\partial x} + \beta k_1 \frac{\partial}{\partial y}\right)^2 f + \cdots\right] \\ &= h\left[f + hD_1 f + \frac{1}{2!} h^2 D_1^2 f + \frac{1}{3!} h^3 D_1^3 f + \cdots\right] \end{aligned} \tag{1.222}$$

having substituted Eq. 1.217 for k_1 and where

$$D_1 \equiv \alpha \frac{\partial}{\partial x} + \beta f \frac{\partial}{\partial y}. \tag{1.223}$$

When the Taylor series in two variables is used to expand k_3 we obtain

$$\begin{aligned} k_3 &= h\left[f(x_0, y_0) + \left(\alpha_1 h \frac{\partial}{\partial x} + \{\beta_1 k_1 + \gamma_2 k_2\} \frac{\partial}{\partial y}\right) f \right. \\ &\quad \left. + \frac{1}{2!}\left(\alpha_1 h \frac{\partial}{\partial x} + \{\beta_1 k_1 + \gamma_2 k_2\} \frac{\partial}{\partial y}\right)^2 f + \cdots\right] \\ &= h\left[f(x_0, y_0) + hD_2 f + h^2 \gamma_1 \left\{D_1 f + \frac{h}{2!} D_1^2 f + \cdots\right\} \frac{\partial f}{\partial y} \right. \\ &\quad \left. + \frac{h^2}{2!} D_2^2 f + h^2 \gamma_1 \{hD_1 f \cdot D_2 f_y + \cdots\}\right] \end{aligned} \tag{1.224}$$

having substituted for k_1 and k_2 with

$$D_2 \equiv \alpha_1 \frac{\partial}{\partial x} + (\beta_1 + \gamma_1) f \frac{\partial}{\partial y}. \tag{1.225}$$

Finally, when we define

$$D_3 \equiv \alpha_2 \frac{\partial}{\partial x} + (\beta_2 + \gamma_2 + \delta_2) f \frac{\partial}{\partial y} \tag{1.226}$$

we can express k_4 as

$$\begin{aligned} k_4 &= h\left[f + hD_3 f + \frac{h^2}{2!} D_3^2 f + \frac{h^3}{3!} D_3^3 f + \cdots \right. \\ &\quad + h^2(\gamma_2 D_1 f + \delta_2 D_2 f) f_y \\ &\quad + \frac{h^3}{2!}(\gamma_2 D_1^2 f + \delta_2 D_2^2 f + 2\gamma_1 \delta_2 f_y D_1 f) f_y \\ &\quad \left. + h^3(\gamma_2 D_1 f + \delta_2 D_2 f) D_3 f_y + \cdots\right] \end{aligned} \tag{1.227}$$

In order that Eqs. 1.216 and 1.221 agree up to h^4 inclusive, the following eight relations must hold among the thirteen parameters

$$
\begin{aligned}
R_1 + R_2 + R_3 + R_4 &= 1 \\
R_2D_1f + R_3D_2f + R_4D_3f &= Df/2! \\
R_2D_1^2f + R_3D_2^2f + R_4D_3^2f &= 2!\, D^2f/3! \\
R_2D_1^3f + R_3D_2^3f + R_4D_3^3f &= 3!\, D^3f/4! \\
R_3\gamma_1D_1f + R_4(\gamma_2D_1f + \delta_2D_2f) &= Df/3! \\
R_3\gamma_1D_1^2f + R_4(\gamma_2D_1^2f + \delta_2D_2^2f) &= 2!\, D^2f/4! \\
R_3\gamma_1D_1fD_2f_y + R_4(\gamma_2D_1f + \delta_2D_2f)D_3f_y &= 3DfDf_y/4! \\
R_4\gamma_1\delta_2D_1f &= Df/4!
\end{aligned}
\tag{1.228}
$$

The operators D_r, $r = 1$, 2, and 3, are defined above. If Eqs. 1.228 are to be independent of f, so that the method applies to all functions f, then the ratios

$$\frac{D_r^jf}{D^jf}$$

must be constant. From the definitions of the D_r operators, these ratios will be constants if

$$\alpha = \beta, \qquad \alpha_1 = \beta_1 + \gamma_1, \qquad \alpha_2 = \beta_2 + \gamma_2 + \delta_2,$$

that is

$$D_1 = \alpha D, \qquad D_2 = \alpha_1 D, \quad \text{and} \quad D_3 = \alpha_2 D.$$

In view of these three relations, the eight equations assume a form independent of $f(x, y)$, namely

$$
\begin{aligned}
R_1 + R_2 + R_3 + R_4 &= 1 \\
R_2\alpha + R_3\alpha_1 + R_4\alpha_2 &= \tfrac{1}{2} \\
R_2\alpha^2 + R_3\alpha_1^2 + R_4\alpha_2^2 &= \tfrac{1}{3} \\
R_2\alpha^3 + R_3\alpha_1^3 + R_4\alpha_2^3 &= \tfrac{1}{4} \\
R_3\alpha\gamma_1 + R_4(\alpha\gamma_2 + \alpha_1\delta_2) &= \tfrac{1}{6} \\
R_3\alpha^2\gamma_1 + R_4(\alpha^2\gamma_2 + \alpha_1^2\delta_2) &= \tfrac{1}{12} \\
R_3\alpha\alpha_1\gamma_1 + R_4(\alpha\gamma_2 + \alpha_1\delta_2)\alpha_2 &= \tfrac{1}{8} \\
R_4\alpha\gamma_1\delta_2 &= \tfrac{1}{24}
\end{aligned}
\tag{1.229}
$$

We now have eight equations in ten unknowns, consequently the system has two degrees of freedom.

To solve this system of equations in terms of α and α_1, we begin by proving $\alpha_2 = 1$. To the fourth we add the second multiplied by $\alpha\alpha_2$ and add the third multiplied by $-(\alpha + \alpha_2)$, then

$$R_3\alpha_1(\alpha - \alpha_1)(\alpha_2 - \alpha_1) = \frac{\alpha\alpha_2}{2} - \frac{\alpha + \alpha_2}{2} + \frac{1}{4} \tag{1.230}$$

From the fifth and seventh it follows that

$$R_3\alpha\gamma_1(\alpha_2 - \alpha_1) = \frac{\alpha_2}{6} - \frac{1}{8}, \tag{1.231}$$

while from the fifth and sixth

$$R_4\alpha_1\delta_2(\alpha_1 - \alpha) = \frac{1}{12} - \frac{\alpha}{6}.$$

R_4 is eliminated between this equation and the eighth to get

$$\alpha\gamma_1 = \frac{\alpha_1(\alpha - \alpha_1)}{2(2\alpha - 1)}.$$

This result is substituted into Eq. 1.231 to give

$$R_3\alpha_1(\alpha - \alpha_1)(\alpha_2 - \alpha_1) = (2\alpha - 1)\left(\frac{\alpha_2}{3} - \frac{1}{4}\right).$$

When this result is compared with Eq. 1.230 we see that their left-hand sides are equal; hence

$$\frac{\alpha\alpha_2}{2} - \frac{\alpha + \alpha_2}{3} + \frac{1}{4} = (2\alpha - 1)\left(\frac{\alpha_2}{3} - \frac{1}{4}\right) = \frac{2\alpha\alpha_2}{3} - \frac{\alpha}{2} - \frac{\alpha_2}{3} + \frac{1}{4}.$$

Hence

$$\alpha(\alpha_2 - 1) = 0,$$

that is

$$\alpha = 0 \quad \text{or} \quad \alpha_2 = 1. \tag{1.232}$$

From the eighth equation it is seen that $\alpha \neq 0$ and $R_4 \neq 0$, therefore we must have $\alpha_2 = 1$. It is evident from Eq. 1.231 that $R_3 \neq 0$.

We now solve the system of equations (1.229) in terms of α and α_1. The first four equations determine R_i uniquely provided the determinant

$$\alpha\alpha_1(\alpha - \alpha_1)(\alpha_1 - 1)(1 - \alpha) \neq 0.$$

The values of the weighting coefficients are

$$R_1 = \frac{1}{2} + \frac{1 - 2(\alpha + \alpha_1)}{12\alpha\alpha_1}$$

$$R_2 = \frac{2\alpha_1 - 1}{12\alpha(\alpha_1 - \alpha)(1 - \alpha)}$$

$$R_3 = \frac{1 - 2\alpha}{12\alpha_1(\alpha_1 - \alpha)(1 - \alpha_1)}$$

$$R_4 = \frac{1}{2} + \frac{2(\alpha + \alpha_1) - 3}{12(1 - \alpha)(1 - \alpha_1)} \tag{1.233}$$

The fifth, sixth, and seventh equations of (1.229) determine γ_1, γ_2, and δ_2 in terms of α and α_1, provided the determinant

$$R_3 R_4^2 \alpha^2 \alpha_1 (\alpha_1 - \alpha)(\alpha_1 - 1) \neq 0.$$

The values are

$$\begin{aligned} \gamma_1 &= \frac{\alpha_1(\alpha_1 - \alpha)}{2\alpha(1 - 2\alpha)} \\ \gamma_2 &= \frac{(1 - \alpha)\{\alpha + \alpha_1 - 1 - (2\alpha_1 - 1)^2\}}{2\alpha(\alpha_1 - \alpha)[6\alpha\alpha_1 - 4(\alpha + \alpha_1) + 3]} \\ \delta_2 &= \frac{(1 - 2\alpha)(1 - \alpha)(1 - \alpha_1)}{\alpha_1(\alpha_1 - \alpha)[6\alpha\alpha_1 - 4(\alpha + \alpha_1) + 3]}. \end{aligned} \tag{1.234}$$

We now have all parameters written in terms of α and α_1. Any two conditions, consistent with the previous equations, may be imposed. For example, if we take

$$R_1 = R_4 \qquad \text{and} \qquad R_2 = R_3, \tag{1.235}$$

this is equivalent to the single condition

$$\alpha + \alpha_1 = 1.$$

To show that the latter condition is consistent, consider

$$\frac{2\alpha_1 - 1}{12\alpha(\alpha_1 - \alpha)(1 - \alpha)} = \frac{1 - 2\alpha}{12\alpha_1(\alpha_1 - \alpha)(1 - \alpha_1)}$$

which leads to the identity

$$\begin{aligned} (2\alpha_1 - 1)(1 - \alpha_1)\alpha_1 &= (1 - 2\alpha)\alpha(1 - \alpha) \\ &= (1 - 2[1 - \alpha_1])(1 - \alpha_1)(1 - [1 - \alpha_1]) \\ &= (2\alpha_1 - 1)(1 - \alpha_1)\alpha_1. \end{aligned}$$

Consequently, the weights and coefficients become

$$\begin{aligned} 12R_1 &= 12R_4 = 6 - \frac{1}{\alpha\alpha_1} \\ 12R_2 &= 12R_3 = \frac{1}{\alpha\alpha_1} \\ \gamma_1 &= \frac{\alpha_1}{2\alpha}, \qquad \gamma_2 = \frac{\alpha_1(\alpha - \alpha_1)}{2\alpha(6\alpha\alpha_1 - 1)}, \qquad \delta_2 = \frac{\alpha}{6\alpha\alpha_1 - 1}. \end{aligned} \tag{1.236}$$

The second condition can be chosen in our values of α and α_1 consistent with their sum being unity; for example,

$$\alpha = \tfrac{1}{3} \qquad \text{and} \qquad \alpha_1 = \tfrac{2}{3}.$$

We then have

$$
\begin{array}{lllll}
R_1 = \frac{1}{8} & & & & \\
R_2 = \frac{3}{8}, & \alpha = \frac{1}{3}, & \beta = \frac{1}{3} & & \\
R_3 = \frac{3}{8}, & \alpha_1 = \frac{2}{3}, & \beta_1 = -\frac{1}{3}, & \gamma_1 = 1 & \\
R_4 = \frac{1}{8}, & \alpha_2 = 1, & \beta_2 = 1, & \gamma_2 = -1, & \delta_2 = 1
\end{array}
\tag{1.237}
$$

which gives Kutta's formulas

$$
\begin{aligned}
k_1 &= hf(x_0, y_0) \\
k_2 &= hf(x_0 + h/3, y_0 + k_1/3) \\
k_3 &= hf(x_0 + 2h/3, y_0 - k_1/3 + k_2) \\
k_4 &= hf(x_0 + h, y_0 + k_1 - k_2 + k_3) \\
\delta y &= \tfrac{1}{8}(k_1 + 3k_2 + 3k_3 + k_4)
\end{aligned}
\tag{1.238}
$$

It is interesting and important to examine the cases when the determinants

$$
\alpha\alpha_1(\alpha - \alpha_1)(\alpha_1 - 1)(1 - \alpha)
$$

$$
\alpha^2\alpha_1(\alpha_1 - \alpha)(\alpha_1 - 1)
$$

vanish. There are only three possible cases in which the solutions are finite:

1. $\alpha = \alpha_1$.
2. $\alpha = 1$.
3. $\alpha_1 = 0$.

For finite R_2 and R_3, case 1 implies the further condition

$$
\alpha = \alpha_1 = \tfrac{1}{2}.
$$

Either R_2 or R_3 may be regarded now as arbitrary, but

$$
R_2 + R_3 = \tfrac{2}{3}.
$$

Let us take

$$
R_2 = R_3 = \tfrac{1}{3},
$$

then

$$
\begin{array}{lllll}
R_1 = \frac{1}{6}, & & & & \\
R_2 = \frac{1}{3}, & \alpha = \frac{1}{2}, & \beta = \frac{1}{2} & & \\
R_3 = \frac{1}{3}, & \alpha_1 = \frac{1}{2}, & \beta_1 = 0, & \gamma_1 = \frac{1}{2} & \\
R_4 = \frac{1}{6}, & \alpha_2 = 1, & \beta_2 = 0, & \gamma_2 = 0, & \delta_2 = 1
\end{array}
$$

which leads to the Runge set of formulas, usually encountered in textbooks,

$$
\begin{aligned}
k_1 &= hf(x_0, y_0) \\
k_2 &= hf(x_0 + h/2, y_0 + k_1/2) \\
k_3 &= hf(x_0 + h/2, y_0 + k_2/2) \\
k_4 &= hf(x_0 + h, y_0 + k_3)
\end{aligned}
\tag{1.239}
$$

with

$$\delta y = \tfrac{1}{6}(k_1 + 2k_2 + 2k_3 + k_4).$$

For the second-order ordinary differential equation

$$\frac{d^2y^{(1)}}{dx^2} = g(x)y^{(1)}$$

we let

$$y^{(2)} \equiv \frac{dy^{(1)}}{dx}.$$

We then have

$$\frac{dy^{(2)}}{dx} = \frac{d^2y^{(1)}}{dx^2} = g(x)y^{(1)}. \tag{1.240}$$

Equations 1.240 are a pair of coupled first-order equations. To each first-order equation there is one arbitrary integration constant to be determined by the boundary conditions. In the section on starting the solution we saw how to determine the derivatives at the boundary. In other words, for the second-order equation we know both $y^1(0)$ and $y^2(0)$, since the latter is merely the first derivative of y^1. For the pair of equations in (1.240) we note that

$$f^{(1)}(x, y^{(1)}, y^{(2)}) = y^{(2)}(x),$$

and

$$f^{(2)}(x, y^{(1)}, y^{(2)}) = g(x)y^{(1)}(x)$$

and so Eqs. 1.239 become

$$
\begin{aligned}
k_1^{(1)} &= hy^{(2)}(x_0); & k_1^{(2)} &= hg(x_0)y^{(1)}(x_0) \\
k_2^{(1)} &= h[y^{(2)}(x_0) + k_1^{(2)}/2]; & k_2^{(2)} &= hg\left(x_0 + \frac{h}{2}\right)[y^{(1)}(x_0) + k_1^{(1)}/2] \\
k_3^{(1)} &= h[y^{(2)}(x_0) + k_2^{(2)}/2]; & k_3^{(2)} &= hg\left(x_0 + \frac{h}{2}\right)[y^{(1)}(x_0) + k_2^{(1)}/2] \\
k_4^{(1)} &= h[y^{(2)}(x_0) + k_3^{(2)}]; & k_4^{(2)} &= hg(x_0 + h)[y^{(1)}(x_0) + k_3^{(1)}].
\end{aligned}
\tag{1.241}
$$

and so we can compute $\delta y^{(1)}$ and $\delta y^{(2)}$ and the whole process repeated over the next step in h.

Problem 11. Write a computer program to calculate the s- and p-wave phase shifts for electrons with incident energies 0.125, 0.25, 0.5, and 1.0 rydberg, scattered by the static field of atomic hydrogen [27].

1.5 PHOTON-ATOM REACTIONS

In Chapter 1 we are attempting to introduce the student to the basic physical notions and theoretical techniques encountered in atomic collision theory. We continue by extending the discussion to the interaction of photons with atomic systems. Here, we are interested in transitions from bound (discrete) states of a target system to a free (continuum) state of the system. When the target is an atom

$$A + h\nu \rightarrow A^{+} + e^{-}$$

then the process is known as photoionization, with the residual system being a positive ion. When the target is a negative ion

$$A^{-} + h\nu \rightarrow A + e^{-}$$

then the process is commonly known as photodetachment. (Finally, when the target system is a nucleus, the process is known as photodissociation.)

When an atom in a *bound* state absorbs a photon, of frequency ν, then there is a probability of the atom making a transition to a *free*, or continuum state. This probability (the cross section) is given by the ratio of the number of photons absorbed per unit time (called the photon transition rate t_ν) to the number of photons in the incident beam crossing unit area in unit time (i.e., the incident photon flux j_ν),

$$\sigma_\nu = \frac{t_\nu}{j_\nu}. \tag{1.242}$$

Absorption processes in which *more* than one photon are involved in ejecting a single electron are rare events, and it is a very good approximation to assert that the number of photons absorbed per unit time equals the number of electrons making bound-free transitions per unit time. Consequently, the electron transition rate, t_e, will be used in place of t_ν in Eq. 1.242. The electron transition rate will depend on the characteristics of the incident radiation, namely its frequency and direction of incidence $\mathbf{k}_\nu$ and polarization $\hat{n}$, and the momentum of the ejected electron, $\mathbf{k}_f$. The differential transition rate is the number of electrons making bound-free transitions into any number of free states with momentum $\mathbf{p}_f = h\mathbf{k}_f$ in the solid angle $d\hat{k}_f$ about $\mathbf{k}_f$

$$dt_e = t_e(\mathbf{k}_f, k_\nu, \hat{n})\, d\hat{k}_f \equiv W_{fi}. \tag{1.243}$$

The solid angle $d\hat{k}_f$ refers to the direction of motion of the ejected electron relative to the direction of incidence of the radiation field. Consequently, the differential cross section is given by

$$d\sigma = \frac{W_{fi}}{j_\nu}. \tag{1.244}$$

This transition rate W_{fi} from initial (bound) to final (free) states can be calculated using first-order time-dependent perturbation theory, using a semi-classical model [28] for the interaction between bound electrons and the radiation field. That is to say, we quantize the motion of the electron but treat the radiation field classically. Consequently, the Hamiltonian for an electron in a center of force due to a potential $V(r)$ and an external electromagnetic field $\mathbf{A}$ is

$$H = \frac{1}{2m}\left(-i\hbar\boldsymbol{\nabla} - \frac{e}{c}\mathbf{A}\right)^2 + V(r) \approx H_0 + H' \tag{1.245}$$

where the 'unperturbed' part of the Hamiltonian is

$$H_0 = -\frac{\hbar}{2m}\nabla^2 + V(r)$$

and the perturbing interaction to first order in $\mathbf{A}$ is

$$H' = \frac{ie\hbar}{mc}\mathbf{A}\cdot\boldsymbol{\nabla}. \tag{1.246}$$

A particular application of this Hamiltonian is the absorption of photons by hydrogen atoms where $V(r)$ is defined in Problem 8. Consequently, in the final state we have the ejected electron moving in the Coulomb field of the residual positive ion (proton).

According to Schiff† the transition rate is given by

$$W_{fi} = \frac{2\pi}{\hbar}|\langle f|\,H'\,|i\rangle|^2\,\rho(k_f), \tag{1.247}$$

where $\rho(k_f)$ is the density of final states for an electron with wave vector $\mathbf{k}_f$, $|i\rangle$ and $|f\rangle$ are the initial and final state functions of the electron. *The density of states, $\rho(\mathbf{k}_f)$, depends upon the normalization of the final state wave function.*

It is shown in Schiff‡ that when the momentum eigenfunctions are restricted to a large but finite cubical box of volume L^3, then

$$u_{\mathbf{k}}(\mathbf{r}) = L^{-3/2}\exp i\mathbf{k}\cdot\mathbf{r} \tag{1.248}$$

† See Ref. 2, p. 199, Eq. 29.12.
‡ See p. 49.

and

$$\int u_l(\mathbf{r})^* u_k(\mathbf{r})\, d\mathbf{r} = \delta_{lk} \qquad k_x = \frac{2\pi n_x}{L},$$

and so on, where n_x, n_y, and n_z are positive, or negative integers, or zero. With this normalization of the final-state wave function, Schiff† derives the density of final states to be

$$\rho(k_f) = \frac{mL^3 k_f}{8\pi^3\hbar^2}\, d\hat{k}_f. \tag{1.249}$$

We now have all the ingredients necessary to substitute into Eq. 1.244, except the incident photon flux, j_ν, which is the intensity associated with the incident photon plane wave divided by $h\nu$, which according to Schiff‡ is

$$j_\nu = \frac{\nu^2}{c}\frac{|A_0|^2}{\hbar\nu}. \tag{1.250}$$

When Eqs. 1.246 and 1.249 are substituted into (1.247) and this result and Eq. 1.250 substituted into Eq. 1.244, we obtain

$$d\sigma = \frac{(2\pi)/\hbar\, |\langle f|\,(ie\hbar/mc)\mathbf{A}\cdot\nabla\,|i\rangle|^2 \cdot (mL^3k_f/8\pi^3\hbar^2)\, d\hat{k}_f}{\nu^2\, |A_0|^2\, c\hbar\nu}$$

where (Schiff, p. 249) in the matrix element we have, for photon *absorption*,

$$\mathbf{A} = |A_0|\, \hat{n} \exp i\mathbf{k}_\nu \cdot \mathbf{r}. \tag{1.251}$$

Using this form for $\mathbf{A}$ and reducing the factors leads to the differential cross-section formula

$$d\sigma(\mathbf{k}_f, \mathbf{k}_\nu, \hat{n}) = \frac{e^2 L^3 k_f}{(2\pi)^2 mc\nu}\left|\int d\mathbf{r}\, u_{\mathbf{k}_f}(\mathbf{r})^* e^{i\mathbf{k}_\nu\cdot\mathbf{r}}\hat{n}\cdot\nabla u_i(\mathbf{r})\right|^2 d\hat{k}_f. \tag{1.252}$$

To obtain the total cross section we must *average* this expression over the incident photon directions, $\hat{k}_\nu$, *and* the two possible polarizations, $\hat{n}$, as well as *sum* (integrate) over the final electron directions, $\hat{k}_f$. However, if the radiation is incident along the positive z-axis, then $\mathbf{k}_\nu \cdot \mathbf{r}$ becomes $k_\nu{}^z Z$, and if the radiation is polarized with its electric vector along the x-axis, then $\hat{n}\cdot\nabla$ becomes $\partial/\partial x$, and if the final state wave function can be represented to sufficient accuracy by the plane wave in Eq. 1.248, then the averaging

† See p. 200.
‡ See p. 248.

procedures are unnecessary and

$$d\sigma(\mathbf{k}_f, k_\nu^z, \hat{n}_x) = \frac{e^2 k_f}{(2\pi)^2 mc\nu} \left| \int d\mathbf{r} e^{-i\mathbf{k}_f\cdot r + i\mathbf{k}_\nu \cdot \mathbf{r}} \frac{\partial}{\partial x} u_i(\mathbf{r}) \right|^2 d\hat{k}_f$$

$$= \frac{e^2 k_f (k_f^x)^2}{(2\pi)^2 mc\nu} \left| \int d\mathbf{r} u_i(r) e^{i\mathbf{K}\cdot\mathbf{r}} \right|^2 d\hat{k}_f, \tag{1.253}$$

having integrated by parts and where

$$\mathbf{K} \equiv k_\nu \hat{z} - \mathbf{k}_f. \tag{1.254}$$

In formula 1.253 we see that our assumption on the polarization of the incident photon beam is manifest by the factor $(k_f^x)^2$. For an *unpolarized* incident beam, which we can take as a linear combination of beams polarized in both the x and y directions, the latter having the factor $(k_f^y)^2$, we merely take the average of the factors, that is $((k_f^x)^2 + (k_f^y)^2)/2$.

Problem 12. Show that the angular distribution of photoelectrons from the ground state of atomic hydrogen is

$$\sigma(\theta, \phi) = \frac{16 e^2 a_0^3 k_f ((k_f^x)^2 + (k_f^y)^2)}{mc\nu(1 + K^2 a_0^2)^4}$$

where a_0 is the Bohr radius and the incident photon beam is unpolarized. Compute angular distributions and total cross sections for photons in the energy region 15(5)100 eV.

In the above formulas we assumed that the motion of the ejected electron in the field of the residual system could be adequately described by a plane wave (i.e., the Born approximation for the scattering of an electron by the bare nucleus); in general we should like to use an improved form of the final-state wave function. For example, for photodetachment, let us write

$$u_\mathbf{k}(\mathbf{r}) = \frac{1}{L^{3/2}} [e^{i\mathbf{k}_f\cdot\mathbf{r}} + v(\mathbf{r})], \tag{1.255}$$

where $v(\mathbf{r})$ represents the scattered part of this eigenstate. In other words, when we consider the scattering of electrons by neutral atoms, see Section 1.1.4, we might expect Eq. 1.255 to have the asymptotic form of a plane wave plus an *outgoing* scattered wave. Indeed if this were the case, then part of the probability amplitude for a photoelectric transition will be associated with electrons going in directions other than that of the plane wave, since all directions are included in the outgoing spherical wave—that is not what the calculation is intended to yield! In other words, our $d\sigma(\hat{k}_f)$ is for a specific $\hat{k}_f$.

Furthermore, part of the probability amplitude for ejection in the direction $\hat{k}_f$ will be included in the calculations for all other plane wave directions, since their associated outgoing waves contribute to the direction $\hat{k}_f$ under consideration. The only way to avoid this inconsistency in the definition of $d\sigma(\hat{k}_f)$ and the improved forms of $u_{\mathbf{k}}(\mathbf{r})$ is to choose the final-state wave function in such a way that it contains no outgoing spherical wave: this is only possible if the wave function is asymptotically a plane wave plus *ingoing* spherical waves.

In the absorption of photons by H^- ions, the final-state wave functions will be obtained from the same problem as electron-hydrogen atom scattering—see Problem 11. Using Eq. 1.30 for the full wave function, $u_{\mathbf{k}}(\mathbf{r})$, we have the following partial wave expansion for the scattered wave

$$v(\mathbf{r}) = \sum_{l=0}^{\infty} \{L^{3/2}\alpha_l R_l(kr) - (2l+1)i^l j_l(kr)\} P_l(\cos\theta),$$

where the r-independent expansion coefficients $\alpha_l(k)$ are determined from the condition that $v(\mathbf{r})$ must be ingoing only! This constant is determined by substituting the asymptotic forms for R_l and j_l and equating to zero the coefficient of $r^{-1}e^{ikr}$:

$$v(\mathbf{r}) = \sum_{l=0}^{\infty} \{L^{3/2}\alpha_l C_l[e^{i(kr-(l\pi/2)+\delta_l)} - e^{-i(kr-(l\pi/2)+\delta_l)}] - (2l+1)i^l[e^{i(kr-(l\pi/2))} - e^{-i(kr-(l\pi/2))}]\}\frac{P_l(\cos\theta)}{2ikr}$$

hence

$$L^{3/2}\alpha_l C_l e^{i\delta_l} - (2l+1)i^l = 0. \tag{1.256}$$

Consequently the asymptotic form of the full wave function, see Eq. 1.30, becomes

$$u_{\mathbf{k}}(\mathbf{r}) \sim \sum_{l=0}^{\infty}(2l+1)i^l e^{-i\delta_l}L^{-3/2}P_l(\cos\theta)\frac{\{e^{i(kr-(l\pi/2)+\delta_l)} - e^{-i(kr-(l\pi/2)+\delta_l)}\}}{2ikr}$$
$$\sim \frac{1}{L^{3/2}}\sum_{l=0}^{\infty}(2l+1)i^l P_l(\cos\theta)(2ikr)^{-1}\{e^{+i(kr-(l\pi/2))} - S^{\dagger}e^{-i(kr-(l\pi/2))}\} \tag{1.257}$$

An alternative form for the full wave function is

$$u_{\mathbf{k}}(\mathbf{r}) \sim \sum_{l=0}^{\infty}(2l+1)i^l e^{-i\delta_l}L^{-3/2}P_l(\cos\theta) \times \frac{\{\sin[kr-(l\pi/2)]\cos\delta_l + \cos[kr-(l\pi/2)]\sin\delta_l\}}{kr}$$
$$\sim L^{-3/2}\sum_{l=0}^{\infty}(2l+1)i^l e^{-i\delta_l}P_l(\cos\theta)\cos\delta_l \times \frac{\{\sin[kr-(l\pi/2)] + R_l(k)\cos[kr-(l\pi/2)]\}}{kr}$$

where $R_l(k)$ is the one-dimensional form of the real and symmetric reactance, or **R**, matrix of the many-channel problem. In solving the radial equation of Problem (8) we impose the asymptotic form

$$F_l(r) \sim \sin\left(kr - \frac{l\pi}{2}\right) + R_l(k)\cos\left(kr - \frac{l\pi}{2}\right) \tag{1.258}$$

that is, unit amplitude for the sine term, then

$$u_{\mathbf{k}}(\mathbf{r}) \sim L^{-3/2} \sum_{l=0}^{\infty} (2l+1) i^l \cos\delta_l e^{-i\delta_l} (kr)^{-1} P_l(\cos\theta) F_l(kr), \tag{1.259}$$

must be substituted into Eq. 1.252 rather than $e^{i\mathbf{k}_f \cdot \mathbf{r}}$. These latter expressions are applicable to photodetachment where the final state is that of a free electron moving in the potential field generated by a neutral atom: for photoionization, the radial equation for F will contain the Coulomb interaction due to the residual ion, in which case $(kr - l\pi/2)$ of Eq. 1.258 is replaced by the Coulomb argument, see Eq. 1.67.

The incident wave lengths for which the absorption effects are important are sufficiently large (~500 Å) compared with the atomic dimensions (~1 Å) that we might expect the dipole approximation, which retains only the first term in the expansion of $e^{i\mathbf{k}_\nu \cdot \mathbf{r}}$ in Eq. 1.252, to be sufficiently accurate, hence

$$d\sigma(\mathbf{k}_f, \mathbf{k}_\nu, \hat{n}) = \frac{e^2 L^3 k_f}{(2\pi)^2 mcv} |\hat{n} \cdot \mathbf{P}|^2 \, dk_f \tag{1.260}$$

where

$$\mathbf{P} \equiv \int d\mathbf{r} u_{\mathbf{k}_f}(\mathbf{r})^* \nabla u_i(\mathbf{r}). \tag{1.261}$$

The above expression for the differential cross section, Eq. 1.260, describes the ionization for a particular $\mathbf{k}_\nu$ and a particular $\hat{n}$, with respect to the axis of quantization. The differential cross section which is observed involves averaging over all the incident radiation directions, k_ν, and polarizations

$$d\sigma(\mathbf{k}_f) = \frac{1}{4\pi} \int dk_\nu \tfrac{1}{2}\{|\hat{n}_1 \cdot \mathbf{P}|^2 + |\hat{n}_2 \cdot \mathbf{P}|^2\} \frac{e^2 L^3 k_f \, d\hat{k}_f}{(2\pi)^2 mcv}, \tag{1.262}$$

where $\hat{n}_1$ and $\hat{n}_2$ are the two polarization directions normal to $\hat{k}_\nu$. We follow Doughty [29] in evaluating this expression. Since $\hat{n}_1$, $\hat{n}_2$, and $\hat{k}_\nu$ form a unit triad, then

$$|\mathbf{P}|^2 = |\hat{n}_1 \cdot \mathbf{P}|^2 + |\hat{n}_2 \cdot \mathbf{P}|^2 + |\hat{k}_\nu \cdot \mathbf{P}|^2$$

and Eq. 1.262 becomes

$$d\sigma(\mathbf{k}_f) = \frac{1}{8\pi} \int dk_\nu \{|\mathbf{P}|^2 - |\hat{k}_\nu \cdot \mathbf{P}|^2\} \frac{e^2 L^3 k_f \, d\hat{k}_f}{(2\pi)^2 mcv}.$$

However, even for a complex vector **P** we have

$$|\hat{k}_\nu \cdot \mathbf{P}|^2 = |\hat{k}_\nu \cdot \hat{P}\,|\mathbf{P}||^2 = |\hat{k}_\nu \cdot \hat{P}|^2\,|\mathbf{P}|^2$$

and so

$$d\sigma(\mathbf{k}_f) = \frac{1}{8\pi}\,|\mathbf{P}|^2 \int d\hat{k}_\nu\{1 - |\hat{k}_\nu \cdot \hat{P}|^2\}\,\frac{e^2L^3k_f\,d\hat{k}_f}{(2\pi)^2mcv}.$$

To evaluate this integral we take $\hat{P}$ as the axis of a set of polar coordinates with $\hat{k}_\nu$ as the radial vector, and so

$$\frac{1}{8\pi}\int d\hat{k}_\nu(1 - |\hat{k}_\nu \cdot \hat{P}|^2) = \frac{1}{8\pi}\int d\hat{k}_\nu(1 - \cos^2\theta) = \tfrac{1}{3}$$

which finally yields

$$d\sigma(\mathbf{k}_f) = \frac{e^2L^3k_f\,d\hat{k}_f\,|\mathbf{P}|^2}{12\pi^2mcv}, \tag{1.263}$$

where **P** is defined in Eq. 1.261 with $u_{\mathbf{k}}(\mathbf{r})$ having the normalization specified in Eq. 1.259.

To reduce Eq. 1.263 into a form suitable for calculation, we shall rewrite the final-state function by expanding $P_l(\hat{k}_f \cdot \hat{r})$, using the addition theorem, see Eq. 1.29,

$$u_{\mathbf{k}_f}(\mathbf{r}) = 4\pi L^{-3/2}\sum_{l,m} i^l e^{-i\delta_l}Y_{lm}(\hat{k}_f)^*G_{klm}(\mathbf{r}) \tag{1.264}$$

where

$$G_{klm}(\mathbf{r}) \equiv \frac{\cos\delta_l(k)\,Y_{lm}(\hat{r})F_l(kr)}{kr}.$$

We recall that the numerical values for the real function $F_l(kr)$ are obtained by the methods of the preceding section, while $\delta_l(k)$ is the phase shift also computed in the scattering problem.

Substituting Eq. 1.264 into Eq. 1.261 gives

$$\mathbf{P} = 4\pi L^{-3/2}\sum_{lm}(-i)^l e^{i\delta_l}Y_{lm}(\hat{k})\int d\mathbf{r}G_{klm}(\mathbf{r})^*\nabla u_i(\mathbf{r})$$

and so the factor in Eq. 1.263 becomes

$$|\mathbf{P}|^2 = \mathbf{P}^* \cdot \mathbf{P} = (4\pi)^2L^{-3}\sum_{lm}\sum_{l'm'} i^{l-l'}e^{-i(\delta_l-\delta_{l'})}Y_{lm}(\hat{k}_f)^*Y_{l'm'}(\hat{k}_f)\mathbf{M}^*_{klm,i}\mathbf{M}_{kl'm',i}$$

where

$$\mathbf{M}_{klm,i} = \int d\mathbf{r}G_{klm}(\mathbf{r})^*\nabla u_i(\mathbf{r}).$$

We shall introduce the theorem for combining two surface harmonics of the same angles

$$Y_{lm}(\hat{k})Y_{l'm'}(\hat{k}) = \sum_{LM}\left[\frac{(2l+1)(2l'+1)}{4\pi(2L+1)}\right]^{1/2}(ll'mm' \mid LM)(ll'00 \mid L0)\,Y_{LM}(\hat{k}) \tag{1.265}$$

where $(ab\alpha\beta \mid c\gamma)$ are Clebsch-Gordan coefficients to be discussed in detail in Section 2.2.

Substituting the above results into Eq. 1.263 gives the differential cross section for a photoelectron being ejected along $\hat{k}_f$,

$$d\sigma(\mathbf{k}_f) = \frac{4e^2}{3mcv}\,k_f\,d\hat{k}_f \sum_{lml'm'LM} i^{l-l'}e^{-i(\delta_l-\delta_{l'})}\left[\frac{(2l+1)(2l'+1)}{4\pi(2L+1)}\right]^{1/2}$$
$$\times\,(ll'mm' \mid LM)(ll'00 \mid L0)\,Y_{LM}(\hat{k}_f)\mathbf{M}^*_{klm,i}\cdot\mathbf{M}_{kl'm',i}. \tag{1.266}$$

The evaluation of the matrix element $\mathbf{M}$ is assigned to the reader as Problem 25 and we see there that the Clebsch-Gordan coefficients

$$(l_i1m_i\mu \mid lm) \qquad \text{and} \qquad (l_i1m_i\mu \mid l'm')$$

are introduced, where l_im_i are the orbital quantum numbers of the initial state. All the magnetic quantum numbers appear in Clebsch-Gordan coefficients, which allows us to sum over m and m', according to the techniques to be described later in Section 2.2, to obtain

$$(1\ 1\ \mu - \mu \mid L0)$$

with $L = 0$ and 2 only!

Two additional forms for the fundamental matrix element, Eq. 1.260, have been given by Chandrasekhar [30]. All three forms are equivalent provided ψ_i and ψ_f are *exact* eigenstates. We note that the form given in Eq. 1.261 is a matrix element of the velocity operator and cross sections computed with this form are said to be the "dipole velocity" values; the other two forms are referred to as the "dipole length" and "dipole acceleration."

Let H be the Hamiltonian for the system, with eigenenergies E_i and E_f, then

$$H\psi_{i,f} = E_{i,f}\psi_{i,f}.$$

We seek the relations between the matrix elements

$$(\psi_f|\,\mathbf{r}\,|\psi_i), \qquad (\psi_f|\,\frac{d}{dt}\,\mathbf{r}\,|\psi_i) \qquad \text{and} \qquad (\psi_f|\,\frac{d^2}{dt^2}\,\mathbf{r}\,|\psi_i)$$

which are obtained using

$$\frac{dA}{dt} = \frac{1}{i\hbar}(AH - HA)$$

for a quantum-mechanical operator which does not depend on the time explicitly. Starting with the acceleration form of the matrix element, we have

$$(\psi_f|\frac{d\mathbf{v}}{dt}|\psi_i) = \frac{1}{i\hbar}(\psi_f|\mathbf{v}H - H\mathbf{v}|\psi_i)$$

$$= \frac{1}{i\hbar}\{E_i(\psi_f|\mathbf{v}|\psi_i) - E_f(\psi_f|\mathbf{v}|\psi_i)\}$$

$$= \frac{E_i - E_f}{i\hbar}(\psi_f|\mathbf{v}|\psi_i), \quad (1.267)$$

while

$$(\psi_f|\frac{d\mathbf{r}}{dt}|\psi_i) = \frac{E_i - E_f}{i\hbar}(\psi_f|\mathbf{r}|\psi_i).$$

These alternative formulas for evaluating the photon absorption coefficient were given by Chandrasekhar when the quantum-mechanical forms for the operators are used, that is, for the dipole velocity

$$\frac{d\mathbf{r}}{dt} = \frac{\mathbf{p}}{m} = \frac{i\hbar}{m}\boldsymbol{\nabla}.$$

For the dipole acceleration operator, we know that classically the force on a particle is related to the gradient of the potential field

$$\mathbf{F} = m\mathbf{a} = -\boldsymbol{\nabla}V \quad \text{or} \quad \frac{d^2\mathbf{r}}{dt^2} = -\frac{1}{m}\boldsymbol{\nabla}V$$

The three alternative forms for the matrix elements then permit Eq. 1.217 to be written in three basic forms

$$d\sigma^v(\mathbf{k}_f) = \frac{e^2L^3k_f\,d\hat{k}_f}{12\pi^2mcv}\left|\int d\mathbf{r}u_{\mathbf{k}_f}(\mathbf{r})\,\boldsymbol{\nabla}u_i(\mathbf{r})\right|^2 \quad (1.268)$$

$$d\sigma^L(\mathbf{k}_f) = \frac{e^2L^3k_f\,d\hat{k}_f mv}{12\pi^2c\hbar^2}\left|\int d\mathbf{r}u_{\mathbf{k}_f}(\mathbf{r})\mathbf{r}u_i(\mathbf{r})\right|^2 \quad (1.269)$$

$$d\sigma^A(\mathbf{k}_f) = \frac{e^2L^3k_f\,d\hat{k}_f}{12\pi^2mcv^3\hbar^2}\left|\int d\mathbf{r}u_{\mathbf{k}_f}(\mathbf{r})\,\boldsymbol{\nabla}Vu_i(\mathbf{r})\right|^2 \quad (1.270)$$

According to Doughty and Fraser [31] extra care must be exercised in defining the potential, V, used in Eq. 1.270; the potential used in this equation must also be used to obtain the approximate wave functions u_i and u_f.

The inverse process to the photoelectric effect involves the capture of an impinging projectile by a target system, with the *emission* of a photon. This is called *radiative recombination*, that is to say a free-bound transition.

According to Schiff,† the probability of finding the system in a final state i, which has an energy lower than the initial state f by about $h\nu$, is proportional to $|\langle i| H'' |f\rangle|^2$, where

$$\langle i| H'' |f\rangle = \frac{-ie\hbar}{mc} \int u_i(r)^* e^{-i\mathbf{k}_\nu \cdot \mathbf{r}} \mathbf{A}_0^* \cdot \nabla u_{\mathbf{k}_f}(\mathbf{r})\, d\mathbf{r} \tag{1.271}$$

in the matrix element for the *emission* of a photon in the direction $\mathbf{k}_\nu$ with polarization $\mathbf{A}_0$.

In order to derive the cross section for radiative recombination, we make use of the detailed balance theorem for inverse processes. In Eq. 1.200 we have the transition probability from state i to another state f; a corresponding formula holds for the inverse process $f \to i$. Furthermore, we see by inspection that

$$\langle f| H' |i\rangle = \langle i| H' |f\rangle^* \tag{1.272}$$

Consequently we obtain the fundamental relation between the transition probabilities

$$\frac{W_{f\leftarrow i}}{\rho(k_f)} = \frac{W_{i\leftarrow f}}{\rho(k_i)} \tag{1.273}$$

Since the cross sections are obtained from the transition probabilities by dividing by the incident flux, then

$$\frac{d\sigma_{f\leftarrow i}}{j_f \rho(k_f)} = \frac{d\sigma_{i\leftarrow f}}{j_i \rho(k_i)}, \tag{1.274}$$

where j_i and $\rho(k_f)$ are given in Eqs. 1.250 and 1.249, respectively.

In Eq. 1.272 we must use the same electron wave function in the photoionization process

$$h\nu + A \to e^- + A^+,$$

as in the inverse radiative capture process

$$e^- + A^+ \to h\nu + A,$$

which was normalized to unit amplitude in Eq. 1.248, if we quantize in a "box" of unit volume. A consequence of this result is that the electron flux in the inverse process would be $j_f = v_f$ electrons crossing unit area in unit time.

In order to treat light quanta and particles at the same time we use the relativistic relations between energy, momentum, and speed. Since

$$\rho(k_i) = (h)^{-3} p^2 \frac{dp}{dE}\, dk_i, \qquad p = \frac{h\nu}{c}, \qquad \frac{dp}{dE} = \frac{1}{c}$$

$$= \nu^2 (hc^3)^{-1}\, d\hat{k}_1.$$

† See p. 249.

Finally, the differential cross section for radiative capture is given by

$$\begin{aligned} d\sigma_{i\leftarrow f} &= \left[\frac{2\pi\nu\,|A_0|^2}{ch}\right]\left[\frac{\nu^2}{hc^3}\right]\left[\frac{1}{v_f}\right]\left[\frac{2\pi h^2}{mk_f}\right] d\sigma_{f\leftarrow i} \\ &= \left[\frac{2\pi\nu}{c}\right]^3 \frac{|A_0|^2}{hk_{f^2}}\, d\sigma_{f\leftarrow i} \end{aligned} \tag{1.275}$$

2

Many-Channel Problems

2.1 EIGENFUNCTION EXPANSION METHOD

2.1.1. Open and Closed Channels

In order to calculate the cross section for *any* scattering process within the framework of nonrelativistic wave mechanics, it is necessary to approximate the Schrödinger equation. One of the most useful approximation schemes, often referred to as the close-coupling approximation, is to expand the overall wave function of the system, consisting of projectile plus target, in terms of the complete set of (assumed known) eigenstates of the target Hamiltonian. In numerical calculations only a few of the lower stationary, or bound, states of the target are retained in the expansion. For example, the most elaborate calculations performed to date on electron-atomic hydrogen-like targets have only computed with at most the $1s$-$2s$-$2p$-$3s$-$3p$-$3d$ states coupled together, although the computer program itself would allow an arbitrary large number of discrete states to be coupled together. The same limitation has been the case in both electron-helium atom scattering, where the code in this case included the target states 1^1S-2^1S-2^3S-2^1P-2^3P and electron-$(1s^22s^22p^q)$ ions, where $0 \leq q \leq 6$ and only the ground state terms are included. For all the practical limitations of the eigenfunction expansion method, many interesting predictions have been made with it.

The expansion coefficients, which are functions of all those variables except those of the target, are unknown at the outset and they turn out to be solutions of coupled second-order ordinary (usually integro-) differential equations with prescribed boundary conditions. These solutions, to be denoted by $F_{\mu}^{(\nu)}(r)$, are interpreted as describing the radial motion of the projectile relative to the target system in the various quantum states. The suffix μ labels the solution vector of the M coupled system of ordinary differential equations, $\mu = 1, 2, 3, \ldots, M$, while the superscript ν denotes that

n_a families of solution vectors which may be extracted from the system of equations, $\nu = 1, 2, 3, \ldots, n_a$. If there are M such expansion coefficients, then the Schrödinger equation has been replaced by an M-channel problem in one-dimension, r. Those channels in which $F_\mu(r)$ oscillates as $r \to \infty$, which belong to radial equations with $k_\mu^2 > 0$, are said to be *open* channels, and we show that there are n_a of them. The remaining channels, ($n_b = M - n_a$) in number, in which $F_\mu(r)$ vanishes exponentially as $r \to \infty$, and belong to radial equations with $k_\mu^2 < 0$, are called *closed* channels. We have seen in Eq. 1.136 that the energy of the projectile, moving relative to the target in a state described by a collection of quantum numbers included in μ, is given by

$$\frac{\hbar^2}{2m} k_\mu^2 = E - E_\mu; \qquad k_\nu^2 = k_\mu^2 + \frac{2m}{\hbar^2}(E_\mu - E_\nu) \tag{2.1}$$

where E_μ is the energy of one of the states in the eigenfunction expansion. In the figure we have drawn a schematic diagram of a target's energy levels, and the energy of the projectile incident on the ground state, k_1^2. The target in Fig. 24 is He^+ with a $1s$ ground state and the degenerate $2l$ as the first excited states.

From both Eq. 2.1 and the diagram we see that at the energy denoted by the horizontal arrow, k_1^2 is positive, but k_2^2, $k_3^2, \ldots$ are negative. In other words, the projectile in this case does *not* have enough energy to excite the level ($2l$); the level is energetically inaccessible. However, $2l$, $3l$, and so on

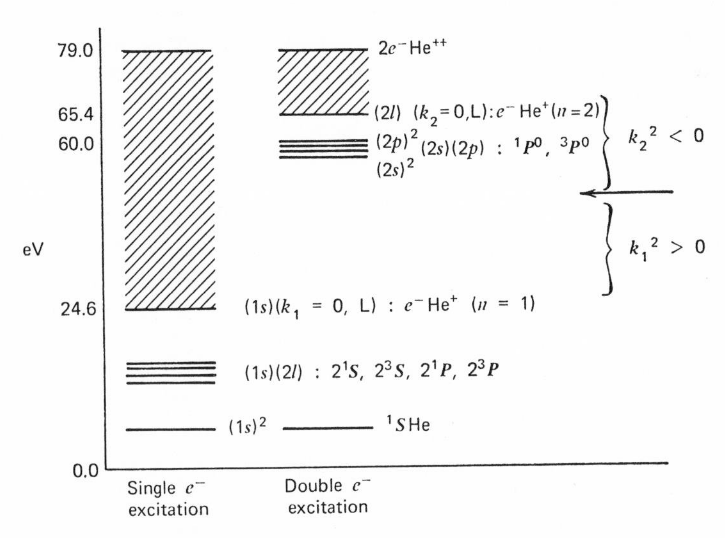

Figure 24 Schematic diagram of the energy levels of singly and doubly excited helium atoms, relative to He ($1s^2$).

may still be included in the mathematical expansion and we speak of virtual transitions to these energetically inaccessible levels.

In general, the number of channels, M, is greater than the number of target states retained in the eigenfunction expansion! This is explained in the following example for electron scattering by atomic systems. The total Hamiltonian for an electron colliding with an N-electron atomic system, neglecting spin-orbit and like interactions, is

$$H_{N+1} = \sum_{i=1}^{N+1} H_1(i) + \sum_{i>j} \frac{e^2}{r_{ij}}, \tag{2.2}$$

where the single-electron Hamiltonian is

$$H_1(i) = -\frac{\hbar^2}{2m}\nabla_i^2 - \frac{Ze^2}{r_i}, \tag{2.3}$$

where e is the electronic charge, Z the number of positive charges at the atomic nucleus, m the electron mass, and $\hbar$ is Planck's constant divided by 2π. This Hamiltonian, H_{N+1}, commutes with the parity operator, the total orbital angular momentum operator, and its Z-component and the total spin operator and its Z-component. Consequently, it is possible to choose a wave function which is simultaneously an eigenfunction of all five operators. In other words, we can choose a representation for the overall wave function in which H is diagonal in $LSM_LM_S\pi$. If several terms of the target atomic system are going to be included, labelled γ_T, then the quantum numbers of the electron colliding with the term T are $k_Tl_T\frac{1}{2}$. The state of the total system is described by specifying the complete set of quantum numbers

$$kl\tfrac{1}{2}, \gamma, LSM_LM_S\pi \equiv \Gamma, \tag{2.4}$$

where $(kl\frac{1}{2}\gamma)$ will be labelled collectively by μ, the channel index.

If we compute the matrix element $(\Gamma|\, H_{N+1}\, |\Gamma')$, where Γ' is an alternative collection of quantum numbers for the system, then we must have the factors $\delta_{LL'}\ \delta_{SS'}\ \delta_{\pi\pi'}\ \delta_{M_LM_L'}\ \delta_{M_SM_S'}$ due to $\psi(\Gamma)$ being a simultaneous eigenfunction of H_{N+1}, L, S, π, L_Z, S_Z. That is to say, each set of quantum numbers $LS\pi M_LM_S$ is not mixed with different sets, and the nonzero matrix elements of H_{N+1} occur in square blocks running down the diagonal. If, for a given set of $LS\pi M_LM_S$, there are M different possible combinations of the set μ, then the corresponding block is $M \times M$. For a given L_T we have that the orbital quantum number of the projectile runs over the range

$$l_T = |L - L_T|, |L - L_T + 1|, \ldots, L + L_T, \tag{2.5}$$

where L_T, as well as $M_{L_T}S_TM_{S_T}$, are elements of γ_T.

Table 3. Projectile Orbital Angular Momenta, l_μ, and Parity of the Total System for H and Ne^+ Systems

Electron-Hydrogen-Like			Electron-Ne^+		
Target State	l_μ	Parity	Target State	l_μ	Parity
$1s$	L	$(-1)^L$	$2p^5\,{}^2P$	$\lvert L-1\rvert, L+1$	$(-1)^L$
				L	$(-1)^{L+1}$
$2s$	L	$(-1)^L$	$2p^6\,{}^2S$	L	$(-1)^L$
$2p$	$\lvert L-1\rvert, L+1$	$(-1)^L$	$3s\,{}^4P$	$\lvert L-1\rvert, L+1$	$(-1)^{L+1}$
	L	$(-1)^{L+1}$		L	$(-1)^L$
$3s$	L	$(-1)^L$	$3s\,{}^2P$	$\lvert L-1\rvert, L+1$	$(-1)^{L+1}$
				L	$(-1)^L$
$3p$	$\lvert L-1\rvert, L+1$	$(-1)^L$	$3p\,{}^4P$	$\lvert L-1\rvert, L+1$	$(-1)^L$
	L	$(-1)^{L+1}$		L	$(-1)^{L+1}$
$3d$	$\lvert L-2\rvert, L, L+2$	$(-1)^L$	$3s\,{}^2D$	$\lvert L-2\rvert, L, L+2$	$(-1)^L$
	$\lvert L-1\rvert, L+1$	$(-1)^{L+1}$		$\lvert L-1\rvert, L+1$	$(-1)^{L+1}$

The various sizes of the blocks can be seen from Table 3, which presents the orbital quantum number l_μ of the projectile for *e-H*-like collisions, and for electrons scattered by Ne^+.

The parity is defined to be $(-1)^{\Sigma_i l_i + l_\mu}$, where i runs over all the orbitals of the target system. So we see that for the $2p$ hydrogen state, where $l_i = 1$, and when $l_\mu = L$, the parity $\pi = (-1)^{L+1}$. Similarly in Ne^+, for the configuration $1s^2 2s^1 2p^6$, we have $\sum_i l_i = 0 + 0 + 6$, and since l_μ can only be L in this case, the parity $\pi = (-1)^{L+6} = (-1)^L$. We have declared parity to be a good quantum number; consequently for a given total orbital angular momentum L, those channels of the system with parity $(-1)^L$ will *NOT* be coupled to those which have parity $(-1)^{L+1}$. For the e^-H-like system the total spin will be either $S = 0$ or 1. We have emphasized that for the Hamiltonian (2.2), S is a good quantum number and so channels belonging to different spin do not mix. In other words, when we fix S and then L, the table shows that once we fix π we can count the number M. For example, for $L \geq 2$ and if we include all states up to $3d$, there will be ten different sets of μ for parity $(-1)^L$, that is, ten different F-functions

$$F_{1sL}, F_{2sL}, F_{2p(L-1)}, F_{2p(L+1)}, F_{3sL}, F_{3p(L-1)}, F_{3p(L+1)}, F_{3d(L-2)},$$
$$F_{3dL}, F_{3d(L+2)}.$$

For $L = 1$, there will be no $|L - 2|$ value of l_μ, and so there will be nine F-functions coupled together; for $L = 0$, there will only be seven F-functions.

We shall remark, somewhat parenthetically here, that He^+ is a hydrogen-like system and so e^-He^+ scattering follows the arguments we have been discussing above and forms a basis for an understanding of the autoionizing states of the helium atom. Both spin states, singlet and triplet, will have the same number of F's.

For e^-Ne^+ system, the total spin $S = 0, 1, 2$ and, as already mentioned, since S is a good quantum number, the set of F's belonging to one $SL\pi$ will not be coupled together with a different set. The e^-Ne^+ system forms a basis for an understanding of the autoionizing states of neon. If we included the two configurations $1s^22s^22p^5$ and $1s^22s2p^6$ only, then for the parity $(-1)^L$, for $L > 0$ we should have three F's, see Table 3.

The eigenfunction expansion method can be regarded as a technique for carrying out configuration interaction in atomic systems. From Table 3 it is clear that we are mixing together target configurations and we are assigning the $(N + 1)$th electron to an l_μ orbital with radial function $F_\mu(r)$. That is to say, we can regard the function $F_\mu(r)$ as a continuum orbital if $k_\mu^2 > 0$ and a discrete orbital if $k_\mu^2 < 0$. Consequently, as we discuss in greater detail later, when we find an autoionizing level of a helium or neon system, say, we can discuss its configuration in terms of the F-orbitals. This is a more

Table 4. Projectile Orbital Angular Momentum l_μ and Parity of the Total System for He and O^{++} Systems

Electron-Helium-Like			Electron-N^+		
Target State	l_μ	Parity	Target State	l_μ	Parity
$ns^2\,{}^1S$	L	$(-1)^L$	$2p^2\,{}^3P$	$\lvert L-1\rvert, L+1$	$(-1)^{L+1}$
				L	$(-1)^L$
$ns(n+1)s\,{}^3S$	L	$(-1)^L$			
			$2p^2\,{}^1D$	$\lvert L-2\rvert, L, L+2$	$(-1)^L$
$ns(n+1)s\,{}^1S$	L	$(-1)^L$		$\lvert L-1\rvert, L+1$	$(-1)^{L+1}$
$ns(n+1)p\,{}^3P$	$\lvert L-1\rvert, L+1$	$(-1)^L$	$2p^2\,{}^1S$	L	$(-1)^L$
	L	$(-1)^{L+1}$			
			$2p^3\,{}^5S$	L	$(-1)^{L+1}$
$ns(n+1)p\,{}^1P$	$\lvert L-1\rvert, L+1$	$(-1)^L$			
	L	$(-1)^{L+1}$	$2p^3\,{}^3D$	$\lvert L-2\rvert, L, L+2$	$(-1)^{L+1}$
				$\lvert L-1\rvert\ L+1$	$(-1)^L$
			$2p^3\,{}^3P$	$\lvert L-1\rvert, L+1$	$(-1)^L$
				L	$(-1)^{L+1}$
			$2p^3\,{}^1D$	$\lvert L-2\rvert, L, L+2$	$(-1)^{L+1}$
				$\lvert L-1\rvert, L+1$	$(-1)^L$
			$3s\,{}^3P$	$\lvert L-1\rvert, L+1$	$(-1)^L$
				L	$(-1)^{L+1}$
			$3s\,{}^1P$	$\lvert L-1\rvert, L+1$	$(-1)^L$
				L	$(-1)^{L+1}$

general way than a variational approach in which an explicit functional form is assumed for the F-orbitals, and when a linear combination of configurations is taken, the arbitrariness is taken up in r-independent coefficients.

In Table 4, the configurations which can be mixed together in the target systems of helium and nitrogen positive ions are presented, along with the projectile's orbital quantum number and the parity of the channels. Since the number of channels is the number of F_μ, we can readily compute the number of channels involved in a computation once we decide how many target states are included in the eigenfunction expansion. Our earlier remark that the number of channels in general exceeds the number of states is clearly illustrated by these tables. For example, associated with the ^3P helium state there are two channels, $l_\mu = |L - 1|$ and $l_\mu = L + 1$ of parity $(-1)^L$. However, these channels will have the same wave number $k^2_{|L-1|} = k^2_{L+1}$. Whether or not a channel is open or closed will depend upon the incident k_1^2, from which we can calculate all the other k_μ^2, using Eq. 2.1 when we know the energies of the target states. Indeed, calculations could be performed with k_1^2 even negative and so all orbitals F_μ are bound state orbitals. However, as explained in detail in Section 1.4.3, we shall only get solutions which are smooth in function and derivative for all r if a true bound state of the system exists at that k_1^2.

2.1.2. Derivation of Radial Equations

In the previous section we have talked about expanding the overall wave function of the target plus projectile system in terms of the eigenfunctions of the target (the motion of the center-of-mass of the system having already been separated off)

$$\Psi(\mathbf{X}, \mathbf{x}_{N+1}) = \sum_{\gamma_T} \psi(\gamma_T, \mathbf{X}) \bar{F}_{\gamma_T}(\mathbf{x}_{N+1}), \tag{2.6}$$

where $\mathbf{X}$ denotes collectively the coordinates of the N electron target, spin as well as spatial, and the sum over the quantum states of the target in practice may include several configurations. The expansion coefficients $\bar{F}$ can be expanded in two steps

$$\bar{F}_{\gamma_T}(\mathbf{x}_{N+1}) = \sum_{m_s} \chi(\tfrac{1}{2}m_s, \sigma_{N+1}) F(\gamma_T m_s, \mathbf{r}_{N+1}) \tag{2.7}$$

where χ is the spin wave function of the projectile, and

$$F(\gamma_T m_s, \mathbf{r}_{N+1}) = \sum_{l_T m_T} Y_{l_T m_T}(\hat{\mathbf{r}}_{N+1}) \frac{f(\gamma_T m_s l_T m_T; r)}{r}. \tag{2.8}$$

When Eqs. 2.7 and 2.8 are substituted back into (2.6) we obtain

$$\Psi(\mathbf{X}, \mathbf{x}_{N+1}) = \sum_{\gamma_T m_s l_T m_T} \psi(\gamma_T, \mathbf{X}) \chi(\tfrac{1}{2}m_s, \sigma_{N+1}) Y_{l_T m_T}(\hat{\mathbf{r}}_{N+1}) \frac{f(\gamma_T m_s l_T m_T; r)}{r}. \tag{2.9}$$

Since the target function $\psi(\gamma_T, \mathbf{X})$ will contain both spin and spatial functions we can combine it with χ and Y, using vector addition coefficients, $(ab\alpha\beta \mid c\gamma)$, discussed in Section 2.2, to give an overall basis function $\psi(\Gamma; \mathbf{X}\hat{\mathbf{x}})$, since

$$\psi(\gamma_T, \mathbf{X})\chi(\tfrac{1}{2}m_s, \sigma_{N+1}) = \sum_{SM_S} (S_T\tfrac{1}{2}M_{S_T}m_s \mid SM_S)\psi(\gamma_T SM_S, \mathbf{X}\sigma_{N+1}),$$

and

$$\psi(\gamma_T SM_S, \mathbf{X}\sigma_{N+1})Y_{l_Tm_T}(\hat{\mathbf{r}}_{N+1}) = \sum_{LM_L} (L_Tl_TM_{L_T}m_T \mid LM_L)\psi(\Gamma; \mathbf{X}\hat{\mathbf{x}}).$$

Substituting these two results back into Eq. 2.9, we sum out the magnetic quantum numbers m_S and m_T by redefining the radial function

$$\begin{aligned} &F(\gamma_T l_T LM_L SM_S; r) \\ &\quad \equiv \sum_{m_s m_T} (S_T\tfrac{1}{2}M_{S_T}m_s \mid SM_S)(L_Tl_TM_{L_T}m_T \mid LM_L)f(\gamma_T m_s l_T m_T; r), \end{aligned} \tag{2.10}$$

and Eq. 2.9 now becomes

$$\Psi(\mathbf{X}, \mathbf{x}_{N+1}) = \sum_{\gamma_T l_T LM_L SM_S = \Gamma} \psi(\Gamma; \mathbf{X}\hat{\mathbf{x}}) \frac{F(\Gamma; r)}{r}. \tag{2.11}$$

Initially, the system has to be in some quantum state, for example when the projectile is incident on the ground state; this fact must be expressed in Eq. 2.11. Consequently, if Γ' denotes the initial complete set of quantum numbers of the overall system, then

$$\Psi(\Gamma'; \mathbf{X}\mathbf{x}_{N+1}) = \sum_{\Gamma} \psi(\Gamma; \mathbf{X}\hat{\mathbf{x}}) \frac{F_\Gamma^{(\Gamma')}(r)}{r}. \tag{2.12}$$

However, we know that when we compute the matrix elements of H_{N+1} with this function we must have $\delta_{LL'}\,\delta_{SS'}\,\delta_{\pi\pi'}\,\delta_{M_LM_L'}\,\delta_{M_SM_S'}$ and so the radial functions need only be characterized by the channel indices, $F_\mu^{(\mu')}$. We have neglected to discuss the effects of electron exchange, since we discuss this in Section 2.4.3.

The simplest way to derive radial equations is to substitute Eq. 2.12 into the Schrödinger equation with the Hamiltonian H_{N+1}, multiply throughout with $\psi(\Gamma''; \mathbf{X}\hat{\mathbf{x}})$, and "integrate" over $\mathbf{X}$ and $\hat{\mathbf{x}}$ to obtain

$$\begin{aligned} &\left[\frac{d^2}{dr^2} - \frac{l_\mu''(l_\mu''+1)}{r^2} + \frac{2mE}{\hbar^2} + \frac{2mZe^2/\hbar^2}{r}\right]F_{\mu''}(r) \\ &\qquad = \sum_{\Gamma} \frac{2mr}{\hbar^2}\int d\mathbf{X}\int d\hat{\mathbf{x}}\,\psi(\Gamma'', \mathbf{X}\hat{\mathbf{x}}) \left| H_N + \sum_{i>j}\frac{e^2}{r_{ij}} \right| \psi(\Gamma; \mathbf{X}\hat{\mathbf{x}}) \frac{F_\mu(r)}{r} \end{aligned}$$

which can be reduced to the form

$$\left[\frac{d^2}{dr^2} + k_\mu^{\,2} - \frac{l_\mu(l_\mu+1)}{r^2} + \frac{2(Z-N)}{r}\right]F_\mu(r) = \sum_{\mu'=1}^{M} V_{\mu\mu'}(r)F_{\mu'}(r). \tag{2.13}$$

Problem 14. In the scattering of electrons by He^+, neglecting the effects of electron exchange, the overall wave function, see Eq. 2.11, can be approximated by the two terms

$$\Psi(r_1, r_2) \approx \psi_{1s}(r_1)F_{1sL}(r_2) + \psi_{2s}(r_1)F_{2sL}(r_2).$$

Derive the radial equations (2.13) for this problem.

2.1.3. The S Matrix

From Eq. 2.13 it is clear that in the asymptotic domain the radial functions are superpositions of ingoing, $e^{-i\theta_\mu}$, and outgoing, $e^{i\theta_\mu}$, spherical waves, where θ_μ has already been defined in Eq. 1.67; consequently

$$F_\Gamma(r) \underset{r\to\infty}{\sim} k_\Gamma^{-1/2}\{A_\Gamma e^{-i\theta_\Gamma} - B_\Gamma e^{i\theta_\Gamma}\}. \tag{2.14}$$

The relation between A_Γ and B_Γ defines the S-matrix, see Blatt and Biedenharn [32]

$$B_\Gamma \equiv \sum_{\Gamma'} S_{\Gamma\Gamma'} A_{\Gamma'} \tag{2.15}$$

where the sum is taken over all the open incident channels. What this relation tells us is that if we know the amplitudes of the incident waves, $A_{\Gamma'}$, the S-matrix determines the amplitudes of the outgoing waves. Substituting the definition (2.15) into (2.14) allows us to define the new radial functions $F_{\Gamma\Gamma'}$ which we have introduced already in Eq. 2.12, because

$$\begin{aligned} F_\Gamma(r) &\sim \sum_{\Gamma'} k_\Gamma^{-1/2} A_{\Gamma'}[\delta_{\Gamma\Gamma'} e^{-i\theta_\Gamma} - S_{\Gamma\Gamma'} e^{i\theta_\Gamma}] \\ &\equiv \sum_{\Gamma'} F_{\Gamma\Gamma'}(r). \end{aligned} \tag{2.16}$$

In numerical calculations it is more convenient to work with real rather than complex numbers, and so

$$\begin{aligned} k_\Gamma^{1/2} A_{\Gamma'}^{-1} F_{\Gamma\Gamma'}(r) &\sim \delta_{\Gamma\Gamma'}\{\cos\theta_\Gamma - i\sin\theta_\Gamma\} - S_{\Gamma\Gamma'}\{\cos\theta_\Gamma + i\sin\theta_\Gamma\} \\ &\sim \sin\theta_\Gamma\{-i\delta_{\Gamma\Gamma'} - iS_{\Gamma\Gamma'}\} + \cos\theta_\Gamma\{\delta_{\Gamma\Gamma'} - S_{\Gamma\Gamma'}\} \end{aligned}$$

which can be rewritten in the form

$$\begin{aligned} \{-i\delta_{\Gamma\Gamma'} - iS_{\Gamma\Gamma'}\}^{-1} A_{\Gamma'}^{-1} F_{\Gamma\Gamma'}(r) \sim k_\Gamma^{-1/2}[\sin\theta_\Gamma &+ \cos\theta_\Gamma\{\delta_{\Gamma\Gamma'} - S_{\Gamma\Gamma'}\} \\ &\times \{-i\delta_{\Gamma\Gamma'} - iS_{\Gamma\Gamma'}\}^{-1}]. \end{aligned}$$

The coefficient of the cosine term is called the reactance or R-matrix and its elements are given by

$$R_{\Gamma\Gamma'} = i(+\delta_{\Gamma\Gamma'} - S_{\Gamma\Gamma'})(\delta_{\Gamma\Gamma'} + S_{\Gamma\Gamma'})^{-1} \tag{2.17}$$

Alternatively, writing **S** in terms of **R** by

$$-iR_{\Gamma\Gamma'}(\delta_{\Gamma\Gamma'} + S_{\Gamma\Gamma'}) = \delta_{\Gamma\Gamma'} - S_{\Gamma\Gamma'}$$

$$S_{\Gamma\Gamma'}(1 - iR_{\Gamma\Gamma'}) = (1 + iR_{\Gamma\Gamma'})\delta_{\Gamma\Gamma'},$$

or

$$\mathbf{S} = (I + i\mathbf{R})(I - i\mathbf{R})^{-1} \tag{2.18}$$

Another important matrix is

$$\mathbf{T} \equiv \mathbf{S} - I.$$

We have discussed the single channel forms of S, R, and T in Chapter 1.

When Eq. 2.16 is substituted into (2.11) we have

$$\Psi(\mathbf{X}, \mathbf{x}_{N+1}) = \sum_{\Gamma\Gamma'} \psi(\Gamma; \mathbf{X}, \hat{\mathbf{x}}) \frac{F_{\Gamma\Gamma'}(r)}{r} \equiv \sum_{\Gamma'} \Psi(\Gamma'; \mathbf{X}\mathbf{x}_{N+1}), \tag{2.19}$$

which was used in Eq. 2.12. The wave functions corresponding to systems initially in either channel Γ_α or Γ_β are given by (see Eqs. 2.12 and 2.16)

$$\Psi(\Gamma_\alpha; \mathbf{X}\mathbf{x}_{N+1}) \sim A_{\Gamma_\alpha} \sum_{\Gamma} \psi(\Gamma; \mathbf{X}\hat{\mathbf{x}}) \frac{1}{k_\Gamma^{1/2} r} \{\delta_{\Gamma\Gamma_\alpha} e^{-i\theta_\Gamma} - S_{\Gamma\Gamma_\alpha} e^{i\theta_\Gamma}\}$$

and

$$\Psi(\Gamma_\beta; \mathbf{X}\mathbf{x}_{N+1}) \sim A_{\Gamma_\beta} \sum_{\Gamma} \psi(\Gamma; \mathbf{X}\hat{\mathbf{x}}) \frac{1}{k_\Gamma^{1/2} r} \{\delta_{\Gamma\Gamma_\beta} e^{-i\theta_\Gamma} - S_{\Gamma\Gamma_\beta} e^{i\theta_\Gamma}\} \tag{2.20}$$

Let us generalize our discussion [32] to the collision between a projectile A, with internal kinetic energy T_A and internal potential energy V_A, and a target B. Let r be the coordinate of relative motion and $V(r)$ the interaction energy between A and B. The Schrödinger equation for such a system is

$$\left[-\frac{\hbar^2}{2m} \nabla_r^{\,2} + T_A + T_B + V_A + V_B - V(r)\right]\Psi(\Gamma_\alpha) = E\Psi(\Gamma_\alpha), \tag{2.21}$$

where the wave function will have the asymptotic form (2.20) but with a different basis. A similar equation to (2.21) applies for the function $\Psi(\Gamma_\beta)$ in which the system is initially in the state Γ_β; in other words, the incoming wave is in channel β. Multiplying (2.21) by $\Psi(\Gamma_\beta)$, and the similar equation for $\Psi(\Gamma_\beta)$ by $\Psi(\Gamma_\alpha)$, subtracting and integrating over all the space of the coordinates r_A, r_B, and r, denoted collectively as τ, gives

$$-\frac{\hbar^2}{2m} \int d\tau \{\Psi_\beta \nabla_r^{\,2} \Psi_\alpha - \Psi_\alpha \nabla_r^{\,2} \Psi_\beta\} + \int d\tau \{\Psi_\beta (T_A + T_B)\Psi_\alpha - \Psi_\alpha (T_A + T_B)\Psi_\beta\} = 0.$$

It has been assumed that V_A, V_B, and V are ordinary functions, not operators, which accounts for terms involving these potential terms canceling out.

Before we can evaluate the second term in the above equation we must describe the basis functions. Since we are assuming both A and B can have some internal structure, then the basis function can be constructed from products of the eigenfunctions of the Hamiltonians of A and B, namely $\phi(r_A)\phi(r_B)$. We are now in a position to apply Green's theorem successively to the r_A and r_B coordinates of the second term. When this is done, the resulting surface integrals over a large sphere are seen to vanish, since $\phi(r_A)$ and $\phi(r_B)$ represent bound state wave functions which tend to zero faster than r^{-1} at large r. Consequently, we are left with

$$\int d\tau\{\Psi'_\beta \nabla_r^2 \Psi'_\alpha - \Psi'_\alpha \nabla_r^2 \Psi'_\beta\} = 0,$$

where Ψ' are defined in Eq. 2.20.

In both terms, the Laplacian does not operate on the ϕ-factors, and so integrating over r_A and r_B reduces the double sum over Γ and Γ' to the diagonal terms $\Gamma = \Gamma'$. The Laplacians only operate on the $F(r)$ part of the wave functions, with asymptotic forms given in Eqs. 2.20. We now apply Green's theorem to the remaining integration over r-space and obtain

$$\lim_{r\to\infty} \sum_\Gamma \{F_\Gamma^{(\Gamma_\beta)}(r)\,\partial_r F_\Gamma^{(\Gamma_\alpha)} - F_\Gamma^{(\Gamma_\alpha)}(r)\,\partial_r F_\Gamma^{(\Gamma_\beta)}(r)\}r^2 = 0.$$

When Eqs. 2.20 are substituted into this result we obtain

$$\sum_\Gamma A_{\Gamma_\alpha} A_{\Gamma_\beta} k_\Gamma^{-1}[\{\delta_{\Gamma\Gamma_\beta} e^{-i\theta_\Gamma} - S_{\Gamma\Gamma_\beta} e^{i\theta_\Gamma}\}\{-ik_\Gamma \delta_{\Gamma\Gamma_\alpha} e^{-i\theta_\Gamma} - ik_\Gamma S_{\Gamma\Gamma_\alpha} e^{i\theta_\Gamma}\}$$
$$- \{\delta_{\Gamma\Gamma_\alpha} e^{-i\theta_\Gamma} - S_{\Gamma\Gamma_\alpha} e^{i\theta_\Gamma}\}\{-ik_\Gamma \delta_{\Gamma\Gamma_\beta} e^{-i\theta_\Gamma} - ik_\Gamma S_{\Gamma\Gamma_\beta} e^{i\theta_\Gamma}\}] = 0.$$

Hence

$$\sum_\Gamma A_{\Gamma_\alpha} A_{\Gamma_\beta} k_\Gamma^{-1}[-ik_{\Gamma_\beta} S_{\Gamma_\beta\Gamma_\alpha} \delta_{\Gamma\Gamma_\beta} + ik_{\Gamma_\alpha} S_{\Gamma_\alpha\Gamma_\beta} \delta_{\Gamma\Gamma_\alpha}]2 = 0,$$

which reduces to

$$2A_{\Gamma_\alpha} A_{\Gamma_\beta} i[-S_{\Gamma_\beta\Gamma_\alpha} + S_{\Gamma_\alpha\Gamma_\beta}] = 0.$$

If this term is to vanish, then we must have

$$S_{\Gamma_\beta\Gamma_\alpha} = S_{\Gamma_\alpha\Gamma_\beta}; \qquad S^T = S \tag{2.22}$$

which proves that $\mathbf{S}$ is a symmetric matrix.

A second property of the **S**-matrix can be derived by premultiplying Eq. 2.21 by Ψ_β^* and the equation conjugate to (2.21) for Ψ_β^* by Ψ_α, and subtracting and integrating over τ we obtain

$$-\frac{\hbar^2}{2m}\int d\tau\{\Psi_\beta^*\,\nabla_r^2\Psi_\alpha - \Psi_\alpha\nabla_r^2\Psi_\beta^*\}$$
$$+\int d\tau\{\Psi_\beta^*(T_A + T_B)\Psi_\alpha - \Psi_\alpha(T_A + T_B)\Psi_\beta^*\} = 0.$$

The second term vanishes as before, while the first term reduces to

$$\sum_\Gamma A_{\Gamma\alpha}A_{\Gamma\beta}{}^*k_\Gamma^{-1}[\{\delta_{\Gamma\Gamma_\beta}e^{i\theta_\Gamma} - S^*_{\Gamma\Gamma_\beta}e^{-i\theta_\Gamma}\}\{-ik_\Gamma\delta_{\Gamma\Gamma_\alpha}e^{-i\theta_\Gamma} - ik_\Gamma S_{\Gamma\Gamma_\alpha}e^{i\theta_\Gamma}\}$$
$$- \{\delta_{\Gamma\Gamma_\alpha}e^{-i\theta_\Gamma} - S_{\Gamma\Gamma_\alpha}e^{i\theta_\Gamma}\}\{ik_\Gamma\delta_{\Gamma\Gamma_\beta}e^{i\theta_\Gamma} + ik_\Gamma S_{\Gamma\Gamma_\beta{}^*}e^{-i\theta_\Gamma}\}] = 0.$$

Various terms cancel in this expression to leave

$$A_{\Gamma_\alpha}A_{\Gamma_\beta{}^*}\sum_\Gamma k_\Gamma^{-1}[-ik_\Gamma\delta_{\Gamma_\alpha\Gamma_\beta}2\delta_{\Gamma\Gamma_\alpha} + ik_\Gamma S_{\Gamma\Gamma_\beta{}^*}S_{\Gamma\Gamma_\alpha}2] = 0,$$

which vanishes, provided that

$$-\delta_{\Gamma_\alpha\Gamma_\beta} + \sum_\Gamma S_{\Gamma\Gamma_\beta{}^*}S_{\Gamma\Gamma_\alpha} = 0$$

or in matrix notation

$$\mathbf{S}^\dagger\mathbf{S} = I \tag{2.23}$$

which proves that the **S** matrix is unitary.

These proofs that the **S** matrix is unitary and symmetrical are not general, because we have assumed that the interactions V_A, V_B, and V are ordinary algebraic functions; this will always be the case in the problems we consider. Nevertheless, the result is indeed a general one; Blatt and Weisskopf [34] show how (1) unitarity is a consequence of the conservation of probability, and (2) the symmetry is a consequence of invariance of the scattering process under time reversal.

In Eq. 2.17 we have defined the **R**-matrix in terms of the elements of the **S**-matrix. We need to know what are the properties of **R** due to **S** being unitary and symmetric. From Eq. 2.17 we have that

$$R_{\beta\alpha}{}^* = -i(\delta_{\alpha\beta} - S_{\beta\alpha}{}^*)(\delta_{\beta\alpha} + S_{\beta\alpha}{}^*)^{-1}$$
$$= -i(S^{*-1}_{\beta\alpha} - \delta_{\alpha\beta})S_{\beta\alpha}{}^*S^{*-1}_{\beta\alpha}(S^{*-1}_{\beta\alpha} + \delta_{\alpha\beta})^{-1}$$

From the unitarity property of the S-matrix we have that

$$S^{*-1}_{\beta\alpha} = S_{\alpha\beta},$$

which allows us to write

$$R_{\beta\alpha}{}^* = i(\delta_{\alpha\beta} - S_{\alpha\beta})(\delta_{\alpha\beta} + S_{\alpha\beta})^{-1} = R_{\alpha\beta} \tag{2.24}$$

which proves that the R-matrix is real and symmetric. This fact is a very valuable aid in checking computer programs.

2.1.4. Autoionization [35]

The electron states of atomic systems are usually classified according to the independent particle model. For example, the electrons of the oxygen atom in the ground configuration are assigned to one-electron orbitals $1s$, $2s$, or $2p$. The total orbital and spin angular momenta for the atom are compounded from the individual electron momenta to give the three allowed terms ^3P, ^1D, and ^1S. The wave functions for the terms will be labelled ϕ for discrete states and ψ for continuum states. These functions, the independent particle wave functions for a many electron system, are eigenstates of some approximate Hamiltonian H_0 of the full Hamiltonian H for the N-electron atomic system. An improved theoretical description of the quantum states of the atomic system, $\Psi(E)$, may be obtained by taking linear combinations of different configurations of the independent particle model. The coefficients, which are r-independent, are chosen so that the $\Psi(E)$ are orthonormal and diagonalize H. Such a procedure is known as configuration interaction theory. Let us emphasize that in this theory all the electrons are assigned to known orbitals, in contrast to the eigenfunction expansion method, where the orbital of the final electron is computed by the method itself. The theory of the interaction of configurations of the independent particle model and its connection with the theory of resonance scattering have been developed by Fano [36]. The predictions of this theory are most dramatic at energies above the first ionization threshold. A schematic diagram of the positions of the energy levels of the helium atom, as predicted by the independent particle model, was presented in Fig. 24.

Consider the sequence of levels based on single-electron excitation: keeping one of the pair of electrons fixed in the $1s$ orbital, energy has to be supplied to excite the other electron first to the $2s$-orbital, more energy being required to excite it to the $2p$-orbital, and so on until this excited electron is free with zero kinetic energy. The energy required to ionize the helium atom, leaving an He^+ ion in the $1s$-state is 24.6 eV. The energy required to excite both electrons to the $(2s)$-orbital is shown as 57.8 eV, while the energy required to ionize the helium atom leaving the He^+ ion in the $n = 2$ quantum state ($2s$ and $2p$ being degenerate in our approximation) is 65.4 eV. Finally, the energy required to remove both electrons and leave them without any kinetic energy is 79.0 eV.

From this figure we see that at 60 eV above the ground state, the independent particle model would predict the existence of two conflicting atomic terms: firstly, it would predict a continuum term ψ with configuration $(1s)(kL)$; secondly, it would predict a stationary eigenstate ϕ of H_0 with configuration

$(2l_1)(2l_1')$. However, when that part of the interaction $H - H_0$, which is neglected in the independent particle model, is taken into account, then ϕ and ψ are superimposed in order to give a more accurate theoretical description of the system. Indeed, one can envisage this coupling of the two independent particle configurations as providing a mechanism for the discrete ϕ to dissociate into an electron (wave number k) and an He^+ ion. This phenomenon is called *autoionization*. Crudely speaking, we can overlap the levels of double-electron excitation, as drawn in Fig. 24, on top of those levels for single-electron excitation, thereby "embedding" the $(2l_1)(2l_1')$ states in the $(1s)(kL)$ continuum.

In other words, the independent particle model, IPM, predicts that at 60.0 eV the two-electron helium system is equivalent either to the scattering of 60.0–24.6 eV electrons by He^+ ($n = 1$) ions or a bound system with both electrons in excited states. Neither description is complete! The existence of the ϕ states embedded in the scattering continuum will allow the collision to proceed via the compound state He*, namely,

$$e^- + He^+ \rightarrow He^* \rightarrow He^+ + e^-$$

such that He* decays with a characteristic lifetime. The existence of the compound state ϕ will manifest itself by a dramatic change in the cross section for $e^-$$He^+$ scattering in the neighborhood of ϕ. Of course the functions ϕ and ψ must have the same total quantum numbers L, S, and π before they can be mixed together. We should expect that there will be a denumerably infinite sequence of structures in the cross section between 57.8 and 65.4 eV due to there being a denumerably infinite sequence of ϕ's as the second electron is found in orbitals with higher and higher principal quantum number, n. In summary, then, when we compute the scattering of electrons by He^+ we must include at least two terms in the eigenfunction expansion, namely,

$$P_{1s}(He^+)F_{1sL} + P_{2l_1}(He^+)F_{2l_1l_2}$$

if we are to reproduce the expected structures between 57.8 and 65.4 eV.

Autoionization is observed in photon absorption spectra at energies beyond the first ionization threshold. From Fig. 24 it is clear that when helium, $1s^2$, absorbs a photon of 60.0 eV energy, both electrons can be excited to $n = 2$ levels. The two-electron system is then said to be quasi-stationary, as it can decay into an $He^+(n = 1)$ ion and a free electron of energy (60.0–24.6) eV. It is to be expected that as the bound electron of He^+ is excited up to higher orbitals, there will be a series limit for each of the observed structures corresponding to all the various ionization thresholds. Once the helium has been doubly excited it can ionize itself by decaying into the nearby continuum.

In Fig. 25 we present an energy level diagram showing continua, shaded blocks, and autoionized levels, dashed lines, for an atomic system with

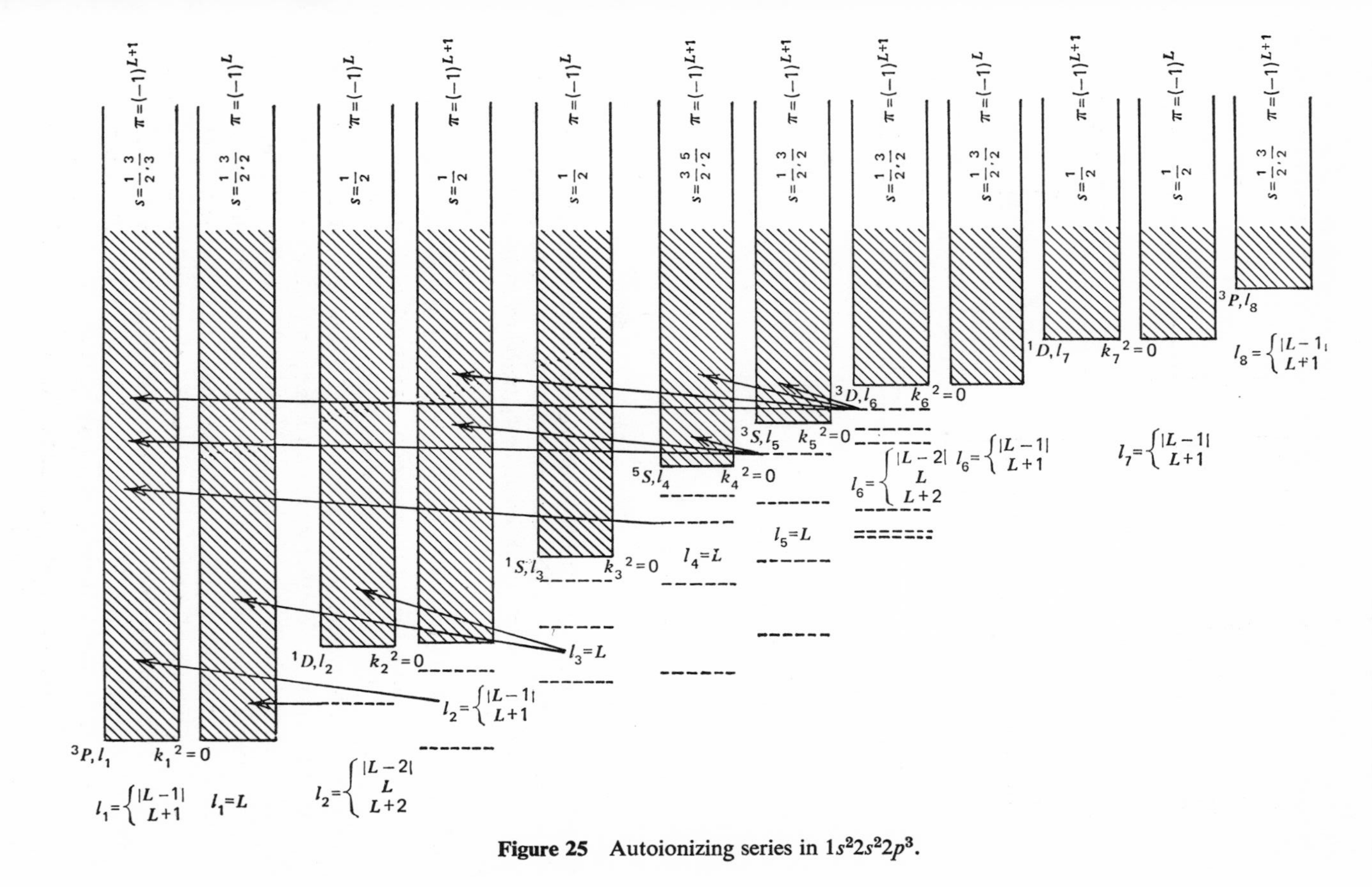

Figure 25 Autoionizing series in $1s^22s^22p^3$.

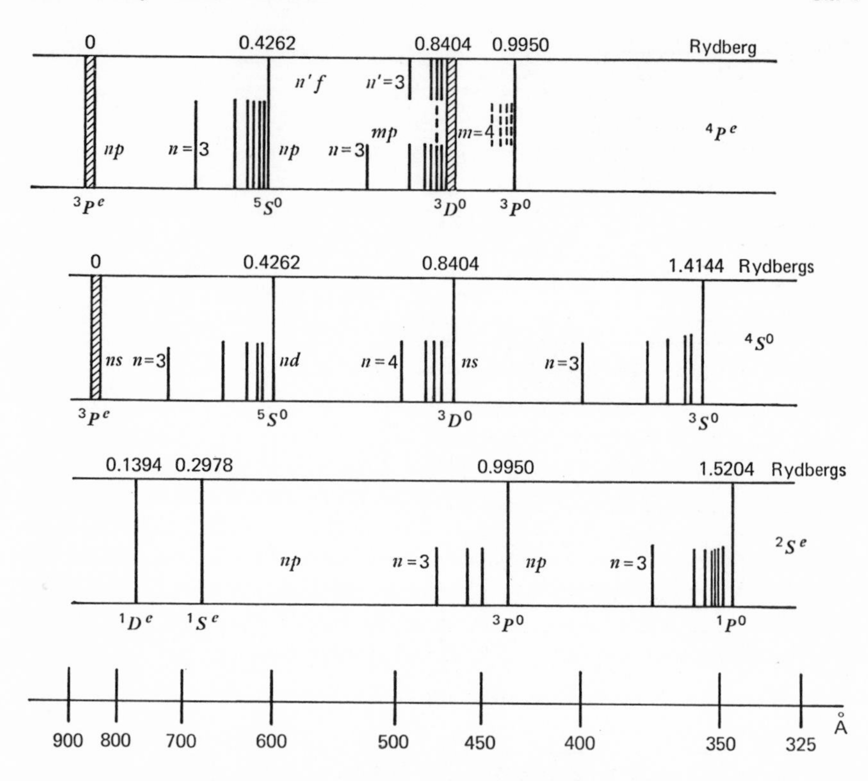

Figure 26 Schematic diagram of the computed spectra from autoionizing states of N in the states $^4P^e$, $^4S^0$, and $^2S^e$. The series limits (identified along the bottom edge of each spectrum) are the atomic terms included in the eigenfunction expansion for N^+.

configuration $1s^22s^22p^3$ for all $LS\pi$. Investigations of the properties of e–N^+, or e–O^{++}, collisions, taking into account the target configurations $1s^22s^22p^2$ and $1s^22s2p^3$ would yield the transitions illustrated in Fig. 25. For a *given* $LS\pi$ for the entire system, the total number of F-orbitals is obtained by counting the number of l_μ. Whether or not an F is open (i.e., a continuum orbital) is determined by whether or not its associated k_μ^2 is positive. The arrows are drawn on Fig. 25 to emphasize into which continua a given autoionizing series can decay. Carroll *et al.* [37] have observed a new Rydberg series in the absorption spectrum of atomic nitrogen which they attributed to transitions from the $^4S^0$ ground state of the nitrogen atom to the Rydberg terms $2s2p^3(^5S^0)np^4P$. This series is one of those shown in Fig. 25, and has been investigated by Smith and Ormonde [38]. In Fig. 26 we show the various autoionizing sequences, and their series limits, as predicted in the calculations of Smith and Ormonde.

2.1.5. Resonance Mechanisms

We now discuss how the eigenfunction expansion method leads to clear predictions of autoionization. Consider Eq. 2.13 for very large values of the radial coordinate: it will be a good approximation to neglect the exponentially decreasing terms in the potential leaving the coupled homogeneous system of equations

$$\frac{d^2F_\mu(r)}{dr^2} \approx \sum_{\gamma=1}^{M}\left[\delta_{\mu\gamma}\left\{\frac{l_\mu(l_\mu+1)}{r^2} - k_\mu^{\ 2} - \frac{2(Z-N)}{r}\right\} + \sum_{\lambda=1}^{\Lambda}\frac{a_{\mu\gamma}{}^{\lambda}}{r^{\lambda+1}}\right]F_\gamma(r). \quad (2.25)$$

The coefficients of the multipole terms, $a_{\mu\gamma}{}^{\lambda}$, which couple together the M channels in the asymptotic region, are readily determined from $V_{\mu\gamma}$, of Eq. 2.13 in a given problem.

For the scattering of electrons by He^+, $Z = 2$ and $N = 1$, which results in an *attractive* Coulomb interaction in all channels. If the incident electron energy $k_1^{\ 2}$ is such that some of the $k_\mu^{\ 2}$, $\mu \neq 1$, are negative, then, neglecting interchannel coupling, the radial equations in these channels have the form of the very familiar bound-state Coulomb problem. That is to say, there will be an infinite number of bound states in each of these channels getting closer together as $k_\mu^{\ 2}$ tends to zero from negative values. Consider Fig. 24, the helium energy level diagram, then the radial functions

$$F_{2sL}, \quad F_{2p|L-1|}, \quad F_{2p|L+1|}$$

belonging to parity $(-1)^L$, will each be bound-state Coulomb wave functions if we have the precise eigenvalues $-|k_2|^2$. In other words, there will be *three* infinite series of bound states. The existence of the coupling terms from these three closed channels to the single open channel, $k_1^{\ 2} > 0$, provides the mechanism by which the bound states decay into the reaction products of the open channel. Of course, a calculation which included higher states, see Table 3, would have more closed channels, and in the decoupled approximation each channel would have an infinite number of bound states. If the coupling potentials between the closed and open channels are weak, then there will be very little incentive for the Coulomb bound states to decay and so they will have very long lifetimes, which is equivalent to a very narrow width, Γ, see Eq. 1.120. Summarizing, we can say that resonances will occur in electron-positive ion collisions due to the attractive Coulomb force in all channels and a many-channel theoretical model will predict their positions and widths.

We can discuss the resonance phenomena in terms of the analytic properties of the **S**-matrix. When we decouple the M channels, the corresponding **S**-matrix is diagonal, $\mathbf{S}_{\alpha\alpha} = e^{i\delta_\alpha(E)}$. The discussion we gave in Section 1.2.1 applies; in the closed channels γ, the bound states will occur as poles on the negative real $k_\gamma^{\ 2}$-axis. For example, when $M = 2$, in other words only

F_{1sL} and F_{2sL} are involved, we can represent the infinite number of poles on the energy plane between the elastic threshold T_1 and the first excitation threshold T_2, crowding upon T_2 as crosses in Fig. 27. At T_μ, we have $k_\mu^2 = 0$. We now want to know the fate of the bound-state poles between T_1 and T_2 when the coupling is turned on. Before the coupling is turned on, the only square-root branch point on the E-plane is at T_2; as soon as there is some coupling a second branch point appears at T_1 and another branch cut should be drawn from T_1 along the real and positive E-axis, thereby generating four energy sheets. Following Eden and Taylor [39], we shall introduce a parameter g to denote the strength of the coupling potential. When $g = 0$, the channels are decoupled. Analytic continuations from P to the bound-state pole can be performed along the two paths A and B. For $g \neq 0$, but as small as we please, a second branch point appears at T_1 and the pole will be displaced slightly from its $g = 0$ position. Each path A and B must lead to a pole! However, there must be *two* poles, because paths A and B no longer terminate at the same point on the *second* sheet of the Riemann surface. The path B terminates at a pole which is very close to the physical energy axis, while the pole at the end of the path A is far from the physical axis. It is the pole B which induces the observed resonant behavior in channel 1, while its shadow pole A will, in general, have no noticeable effect. For M channels coupled together there will be 2^N energy sheets and $(2^N - 1)$ shadow poles. The energy at which the resonance pole lies can be written, in rydbergs, as

$$k_s^2 = E_s - i\frac{\Gamma}{2} \tag{2.26}$$

The same fate will befall all the bound-state poles lying between T_1 and T_2, they will move off the physical energy axis when T_1 is turned on.

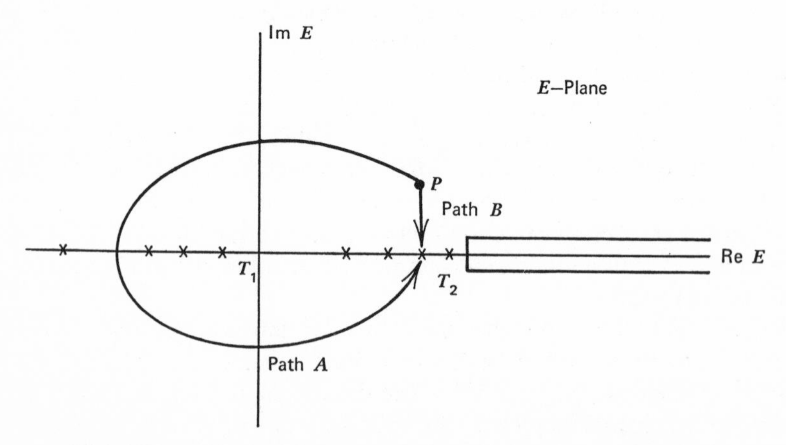

Figure 27 Movement of a bound state pole between two thresholds.

Table 5. Values of the Multipole Index λ for the Potentials $V_{\mu\gamma}$ in Electron-Hydrogen-Like Systems

μ \ γ	$1sL$	$2sL$	$2p\,\|L-1\|$	$2p\,\|L+1\|$
$1sL$	—	—	1	1
$2sL$	—	—	1	1
$2p\,\|L-1\|$	1	1	—	2
$2p\,\|L+1\|$	1	1	2	—

For the scattering of electrons by hydrogen atoms, $Z = 1 = N$, so that there is no residual attractive Coulomb interaction to support the infinite sequence of bound states and thereby generate resonances as described for positive ions. However, the $a_{\mu\gamma}{}^{\lambda}$ in Eq. 2.25 are nonzero for $\mu \neq \gamma$; that is, there are dipole potential terms coupling the channels together. Of course the repulsive centrifugal force appearing in all the diagonal elements of the potential is a dipole term also. The values of λ which appear in the eigenfunction expansion method which retains only the $1s$-$2s$-$2p$ states of H are given in Table 5.

Problem 14. Justify the values quoted in Table 5. What are the values of $a_{\mu\nu}^{(\lambda)}$?

We must emphasize that the coulomb, $2(Z - N)/r$, that is, $\lambda = 0$, and centrifugal forces, $l_{\mu}(l_{\mu} + 1)/r^2$, that is, $\lambda = 2$, are not included in the table. In the approximation keeping the dipole terms connecting the degenerate $2s$-$2p$ channels we show that the resulting potentials in the radial equations once again can support an infinite number of bound states. The dipole coupling to the $1s$ state provides the decay mechanism for the quasi-stationary state and the strength of this coupling determines the width of the associated resonances.

It is probably worthwhile to repeat in somewhat different words what we are trying to do in this section: we know that experimental observations, of a wide variety of phenomena, using a wide range of apparatus, yield cross sections which exhibit a wealth of structure. Our task is to construct a mathematical model to describe all this phenomena from a unified viewpoint, if this is at all possible. We are beginning to see that if we solve the Schrödinger equation using an eigenfunction expansion we are led at once to a many-channel model which seems to have many of the desired qualitative features.

For electrons incident on positive ions we have predicted an infinite set of resonance sequences, each sequence having an infinite number of members. For electrons incident on atomic hydrogen, provided we assume that $2s$ and $2p$ have the same energy, which we know they do *not* have, we shall again predict infinite sequences. Although we must keep in mind that our assumption on the degeneracy of $2s$ and $2p$ may modify this picture.

In order to develop a model for electron-hydrogen atom scattering we follow Gailitis and Damburg [40] and retain only the (4, 2) and (2, 4) elements of Table (5) when $L = 0$. Recall that for $L = 0$ we can only have three coupled second-order ordinary differential equations in (2.25) with solutions labelled $F_{1s(0)}$ $F_{2s(0)}$ $F_{2p(1)}$; furthermore, if we decouple $F_{1s(0)}$ from the other two and $k_1^2 < 0.75$ ryd, then $k_2^2 < 0$ and the pair of coupled equations for $F_{2s(0)}$ and $F_{2p(1)}$ is a bound state problem. For $L > 0$, Gailitis and Damburg retain the four elements (2, 3), (2, 4), (3, 2), and (4, 2) of Table 5, but neglect the quadrupole term in the $2p$ channel, since this goes to zero asymptotically faster than the dipole terms. For $r > r_B$, the radial equation (2.25) reduces to the matrix form

$$\left[\frac{d^2}{dr^2} - \frac{\mathbf{l}(\mathbf{l}+\mathbf{1}) + \mathbf{a}}{r^2} + k^2\right]\mathbf{F}(r) = 0, \tag{2.27}$$

where according to Seaton [41] (see Problem 14) the coefficients of the dipole tails are

$$a_{42} = a_{24} = 6\left[\frac{L}{2L+1}\right]^{1/2}, \qquad a_{32} = a_{23} = -6\left[\frac{L+1}{2L+1}\right]^{1/2} \tag{2.28}$$

The matrix $\{\mathbf{l}(\mathbf{l}+\mathbf{1}) + \mathbf{a}\}$ can be written out as

$$\left|\begin{matrix} L(L+1) & 0 & 0 & 0 \\ 0 & L(L+1) & -6\left[\frac{L+1}{2L+1}\right]^{1/2} & 6\left[\frac{L}{2L+1}\right]^{1/2} \\ 0 & -6\left[\frac{L+1}{2L+1}\right]^{1/2} & |L-1|(|L-1|+1) & 0 \\ 0 & 6\left[\frac{L}{2L+1}\right]^{1/2} & 0 & (L+1)(L+2) \end{matrix}\right| \tag{2.29}$$

and can be diagonalized. In other words, we can set up an eigenvalue problem

$$\{\mathbf{l}(\mathbf{l}+\mathbf{1}) + \mathbf{a}\}\mathbf{A} = \mathbf{A}\boldsymbol{\gamma} \equiv \mathbf{A}\boldsymbol{\mu}(\boldsymbol{\mu}+\mathbf{1}), \tag{2.30}$$

where $\boldsymbol{\gamma} \equiv \boldsymbol{\mu}(\boldsymbol{\mu}+1)$ are the eigenvalues and $\mathbf{A}$ is the matrix of eigenvectors. That is to say, we can diagonalize the numerator of r^{-2} in Eq. 2.27 by premultiplying the equation by $\mathbf{A}^{-1}$ and introducing $\mathbf{AA}^{-1} = \mathbf{I}$ as follows

$$A^{-1}\left[\frac{d^2}{dr^2} - \frac{l(l+1)+a}{r^2} + k^2\right]AA^{-1}F(r) = 0. \tag{2.31}$$

Since the diagonal matrix d^2/dr^2 can be written as the scalar d^2/dr^2 multiplying the identity matrix, which commutes with A, the first term is simply d^2/dr^2. The structure of the matrix given in (2.29) decouples completely the channel F_{1sL} from the other channels and this fact will be reflected in the matrix $\mathbf{A}$—it will *not* mix any of the F_{1sL} with $F_{2l_1l_2}$. In other words, since the 3×3 submatrix of $\mathbf{k}^2$ formed by crossing out the first row and column can be written as the scalar k_2^2 times the unit matrix, I, the structure of A does not mix k_1^2 and k_2^2, and so $\mathbf{A}$ and $\mathbf{k}$ commute. Equation 2.31 becomes

$$\left[\frac{d^2}{dr^2} - \frac{\mu(\mu+1)}{r^2} + k^2\right][A^{-1}F(r)] = 0 \tag{2.32}$$

where all the quantities are diagonal. We can interpret this result as follows: instead of working with the regular set of basis functions in F-space, we use a rotated basis defined by $\mathbf{A}^{-1}\mathbf{F}$, which actually diagonalized the radial equations. Let us remind ourselves that we are considering incident electron energies where $k_2^2 < 0$ and so we have a bound-state problem. The fact that we have been able to decouple our equations into the form (2.32) means that we can use either the method of Section 1.4.3 to extract the bound states, or since the potential is so simple, we can use analytic methods. The appropriate analysis has already been referred to in Problem 6.

Problem 15. Determine the eigenvalues and eigenvectors of the matrix (2.29), for $L = 0, 1, 2, 3$.

If we solve the quadratic $\mu^2 + \mu - \gamma = 0$ we obtain the roots

$$\mu = -\tfrac{1}{2} \pm (\tfrac{1}{4} + \gamma)^{1/2}, \tag{2.33}$$

in terms of the "eigenvalues" γ. For $\gamma < \frac{1}{4}$, $(\mu + \frac{1}{2})$ is imaginary, and Landau and Lifshitz show that the discrete spectra of Eq. 2.32 contains an infinite number of bound states! For $L = 0$, we recall that only $F_{2s(0)}$ and $F_{2p(1)}$ will be allowed and so there will be two eigenvalues which Seaton gives as $\gamma = 1 \pm (37)^{1/2}$. The lower sign fulfills the condition of an infinite number of bound states for $k_2^2 < 0$. In the rotated space, the $\mathbf{S}$-matrix will be diagonal and so one of its diagonal elements will have this infinite sequence of bound-state poles; in the original space, these poles appear in all elements. The fate of these bound-state poles as the coupling to the open F_{1sL} channel was considered above with the aid of Fig. 27: we saw that a corresponding infinite number of resonance poles would be generated. It is usual to denote the motion of the poles, as the coupling is gradually turned on from their position ε_s in the bound-state problem, along and perpendicular to the real energy axis by Δ and Γ respectively. If the coupling is weak, then the poles will not stray too far from the real axis, Γ small, and *each* pole, now on the unphysical sheet of the Riemann surface of $S(E)$, will increase $\delta_L(E)$ in the elastic 1*s* channel by π radians. Since the analysis on the number of bound states in the truncated 2*s*-2*p* problem is independent of the spin, there must be an infinite number of resonances in both ^3S and ^1S. Gailitis and Damburg noted that one of the μ's were imaginary for $L = 1$ and 2; consequently, there are an infinite number of resonances in both the singlet and triplet states of P and D waves.

We can summarize our discussion on resonance mechanisms by observing that attractive Coulomb and dipole potential tails generate bound states in closed channels which subsequently decay into the open channels to which they are coupled. Fonda [42] has shown that, in his model problem, a narrow resonance need have no relation to possible bound states of the decoupled closed channels.

As an explicit example of this (resonances without bound states in closed channels) let us consider the single-channel problem of electrons scattered by nitrogen atoms. The effective potential for this scattering process was presented in Fig. 5. We hinted in Section 1.1.4 that the peak in Fig. 5 has a special role. When the second-order ordinary differential equation is solved, we find the phase shifts, when $LS\Pi = 1$, 1, even, are those drawn in Fig. 28*a*, which gives rise to the resonant form of the total cross section presented in Fig. 28*b*; these results and the subsequent analysis are taken from Henry *et al.* [43].

The rapid increase in the phase shift, although not completely π radians, is due to the "shape" of the potential. Namely, a potential well surrounded by an angular momentum barrier ($l_2 = 1$ in this case). Hence the name "shape resonance." In other words, the total cross section is dominated by the behavior of a particular partial wave. Although the calculations of Henry *et al.* were many-channel in nature, the closed channels do not affect the qualitative result of this resonance. Indeed a rather comprehensive analysis of negative ions and shape resonances has been carried out by Hunt and Moiseiwitsch [44], using the single-channel formalism and a model potential.

If we have a resonance, then the first question we ask is what is its width and position. Since shape resonances in atomic collisions are most often found just above a threshold in an open channel, then it is possible to parametrize the phase shift in that channel. Schwinger's parameterization of the phase shift is known as "effective-range theory" and the fundamental

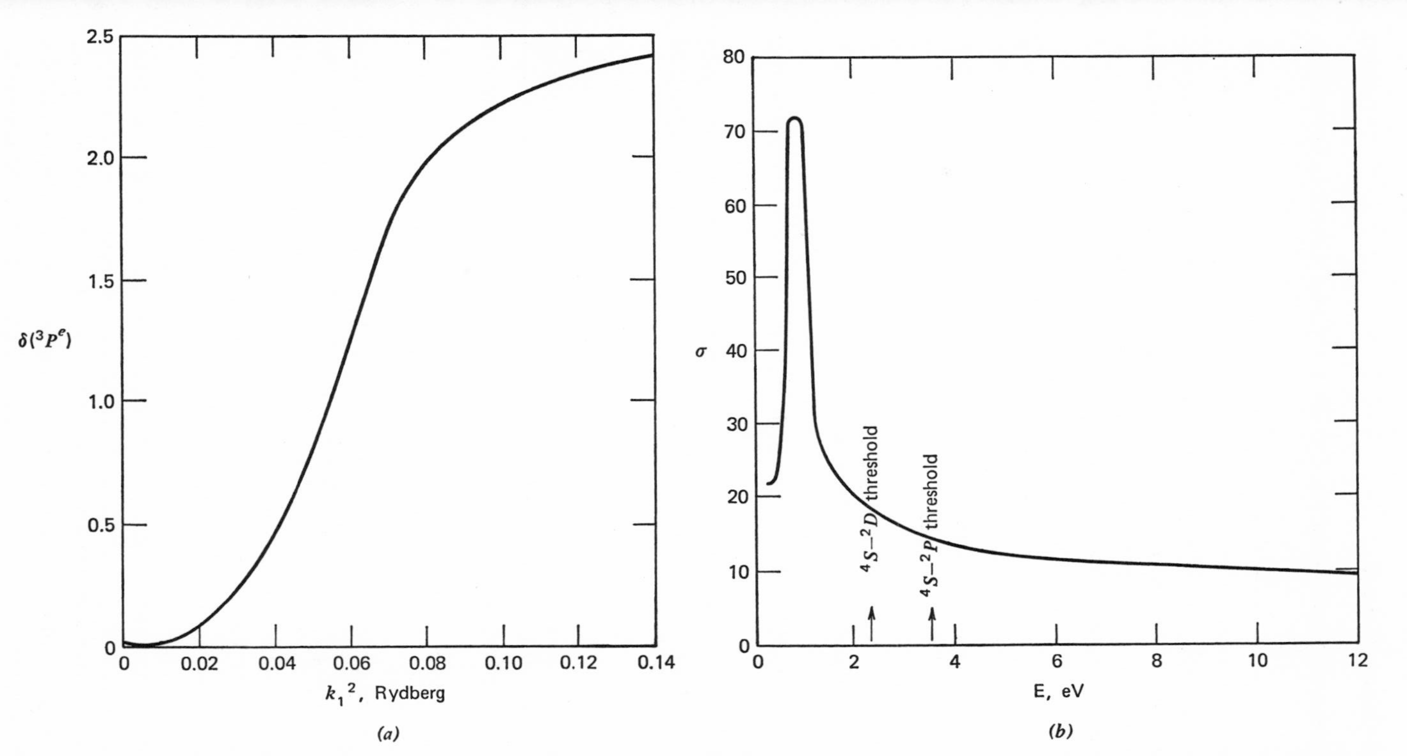

Figure 28 (*a*) The $^3P^e$ phase shift, in radians, versus energy, in Rydbergs, for electron scattering by atomic nitrogen in its lowest atomic term, that is, $^4S^0$. (*b*) Energy dependence, in electron volts, of the total cross sections, in 10^{-16} cm^2, for electrons incident on atomic nitrogen.

formula, for short-range potentials, is

$$k^{2l+1}\cot\delta_l = -\frac{1}{a} + \tfrac{1}{2}r_0k^2, \tag{2.34}$$

where a is called the scattering length and r_0 the effective range. The derivation of Eq. 2.34 is based on straightforward manipulations of the single-channel radial equation, and we refer the reader to Bethe [45].

In Eq. 1.42 we defined the T matrix by

$$T = 2ie^{i\delta_l}\sin\delta_l = \frac{2i}{(\cot\delta_l - i)} \tag{2.35}$$

From Eq. 2.35 we see that T has a pole when $\cot\delta_l = i$; when this result is substituted into Eq. 2.34 we have

$$ik^{2l+1} = -\frac{1}{a} + \tfrac{1}{2}r_0k^2$$

When this equation is compared with Eq. 1.110 we have the position of the resonance given by

$$E_r = \frac{2}{ar_0}, \tag{2.36}$$

and the width by

$$\Gamma = \frac{-4k^{2l+1}}{r_0}.$$

This definition of the width is energy dependent. The usual definition is to define Γ at the resonant energy, hence

$$\Gamma = \frac{-4(2/ar_0)^{l+1/2}}{r_0}. \tag{2.37}$$

We can derive the Breit-Wigner form of the T-matrix in the vicinity of the resonance by substituting Eq. 2.34 into (2.35) to obtain

$$\begin{aligned} T &= \frac{2ik^{2l+1}}{-1/a + \frac{1}{2}r_0k^2 - ik^{2l+1}} \\ &= \frac{i\Gamma}{(E_r - k^2) - i\Gamma/2}. \end{aligned} \tag{2.38}$$

In the e^-N problem, when $k^3\cot\delta$ is plotted versus k^2, we obtain both the scattering length a and the effective range r_0 which leads to $E_r = 0.06577$ ry and $\Gamma = 0.04688$ ry.

2.2 RACAH ALGEBRA

In this section, the algebra of the addition of angular momenta as applied to atomic collision problems is surveyed very briefly. Results and definitions are intermingled; the reader who wants a detailed exposition of this topic is referred to the books of Edmonds [46], Rose [47], and Rotenberg *et al.* [48].

2.2.1. Clebsch-Gordan Coefficients

Let $\mathbf{J}_1$ and $\mathbf{J}_2$ be two angular momentum operators. Their Cartesian components satisfy the commutation relations

$$[J_x, J_y] = iJ_z, \qquad [J^2, J_z] = 0, \tag{2.39}$$

and they commute with each other,

$$[\mathbf{J}_1, \mathbf{J}_2] = 0.$$

For example, $\mathbf{J}_1$ and $\mathbf{J}_2$ may be either the orbital and spin angular momenta of the same particle or they may be the orbital angular momenta of two different particles. We recall a theorem of quantum mechanics that if two operators commute, then the operators can have the same eigenfunction. Let the simultaneous eigenfunctions of J^2 and J_z be $Y_{jm}(\theta\phi)$, then

$$\begin{aligned} \mathbf{J}^2 Y_{jm}(\theta\phi) &= j(j+1)Y_{jm}(\theta\phi) \\ J_z Y_{jm}(\theta\phi) &= mY_{jm}(\theta\phi), \end{aligned} \tag{2.40}$$

where $j(j+1)$ and m are the eigenvalues of the operators J^2 and J_z, respectively. When j is an integer, Y_{jm} are the spherical harmonics of Eq. 1.28; see Condon and Shortley [49].

When we have a product of a pair of angular momentum eigenfunctions we find it convenient to develop a notation for the product, for example,

$$Y_{j_1m_1}(\theta_1\phi_1)Y_{j_2m_2}(\theta_2\phi_2) \equiv |j_1m_1j_2m_2\rangle. \tag{2.41}$$

This product wave function will belong to eigenvalues $j_2(j_2+1)$, m_2, $j_1(j_1+1)$, and m_1. This product wave function will also be an eigenfunction of the z-component of the total angular momentum operator

$$\mathbf{J}_z = \mathbf{J}_{1z} + \mathbf{J}_{2z}$$

with eigenvalue $(m_1 + m_2)$, since

$$\begin{aligned} \mathbf{J}_z|j_1m_1j_2m_2\rangle &= (\mathbf{J}_{1z} + \mathbf{J}_{2z})Y_{j_1m_1}Y_{j_2m_2} \\ &= m_1Y_{j_1m_1}Y_{j_2m_2} + m_2Y_{j_1m_1}Y_{j_2m_2}. \end{aligned}$$

In general, the product wave function is not an eigenfunction of

$$\mathbf{J}^2 = (\mathbf{J}_1 + \mathbf{J}_2)^2,$$

since the eigenfunctions (2.41) are *not* eigenfunctions of $\mathbf{J}_1$ or $\mathbf{J}_2$, due to the commutation relations (2.39).

In order to give a complete quantum mechanical description of a system it is necessary to specify the quantum numbers of a complete set of commuting operators. If $\mathbf{J}_1$, $\mathbf{J}_{1z}$, $\mathbf{J}_2$, and $\mathbf{J}_{2z}$ are members of such a set, let γ denote the remainder of the set needed to specify the system, then we can denote their simultaneous eigenfunctions by

$$|\gamma j_1 j_2 m_1 m_2\rangle.$$

We recall that to every classical degree of freedom there corresponds a quantum number. The question as to whether there is an alternative set of coordinates to those giving rise to $\gamma j_1 j_2 m_1 m_2$ can obviously be answered affirmatively when we generate the set $\gamma j_1 j_2 jm$ of quantum numbers which can label the simultaneous eigenstates

$$|\gamma j_1 j_2 jm\rangle$$

of the set of operators $\boldsymbol{\gamma}$, $\mathbf{J}_1$, $\mathbf{J}_2$, $\mathbf{J}$, $\mathbf{J}_z$. These two sets of eigenstates can be used as bases in which to expand the alternative set, that is,

$$|\gamma j_1 j_2 jm\rangle = \sum_{m_1=-j_1}^{j_1} \sum_{m_2=-j_2}^{j_2} (j_1 j_2 m_1 m_2 \,|\, jm)\, |\gamma j_1 m_1 j_2 m_2\rangle, \tag{2.42}$$

where the expansion coefficients $(j_1 j_2 m_1 m_2 \,|\, jm)$ are called the vector addition or Clebsch-Gordan coefficients; they are real numbers. The algebraic derivation of these coefficients is given in Condon and Shortley [49], and will not be reproduced here. We shall adopt the phase convention of these authors. The two principal properties of the arguments of the coefficients are

$$m = m_1 + m_2,$$

and $(j_1 j_2 j)$ satisfy the triangular inequality: The sum of any pair must be greater than or equal to the third member of the triad.

Let us compute the overlap integral

$$\langle \gamma j_1 j_2 j'm' \,|\, \gamma j_1 j_2 jm\rangle = \sum_{m_1'm_2'} \sum_{m_1 m_2} (j_1 j_2 m_1' m_2' \,|\, j'm')(j_1 j_2 m_1 m_2 \,|\, jm) \times \langle \gamma j_1 m_1' j_2 m_2' \,|\, \gamma j_1 m_1 j_2 m_2\rangle,$$

and we obtain

$$\begin{aligned}\delta_{jj'}\,\delta_{mm'} &= \sum_{m_1'm_2'} \sum_{m_1 m_2} (j_1 j_2 m_1' m_2' \,|\, j'm')(j_1 j_2 m_1 m_2 \,|\, jm\rangle\, \delta_{m_1'm_1}\, \delta_{m_2'm_2} \\ &= \sum_{m_1 m_2} (j_1 j_2 m_1 m_2 \,|\, j'm')(j_1 j_2 m_1 m_2 \,|\, jm),\end{aligned} \tag{2.43}$$

which is an orthonormality property of the vector addition coefficients.

Multiply Eq. 2.42 by $(j_1 j_2 m_1' m_2' \mid jm)$ and sum over j and m to obtain

$$\sum_{jm} (j_1 j_2 m_1' m_2' \mid jm) \, |\gamma j_1 j_2 jm\rangle = \sum_{jm} \sum_{m_1 m_2} (j_1 j_2 m_1' m_2' \mid jm)(j_1 j_2 m_1 m_2 \mid jm)$$

$$= |\gamma j_1 m_1 j_2 m_2\rangle$$

$$= |\gamma j_1 m_1' j_2 m_2'\rangle \tag{2.44}$$

provided the vector addition coefficients satisfy a second orthonormality relation

$$\sum_{jm} (j_1 j_2 m_1' m_2' \mid jm)(j_1 j_2 m_1 m_2 \mid jm) = \delta_{m_1 m_1'} \, \delta_{m_2 m_2'}. \tag{2.45}$$

According to the Wigner-Eckart theorem (see Edmonds, p. 75), the geometrical character of the matrix element $(l'm'|\, Y_{LM} \,|lm)$ can be factored out, namely,

$$(l'm'|\, Y_{LM} \,|lm) = (lLmM \mid l'm')(l' \| \, Y_L \, \| l)(2l' + 1)^{-\frac{1}{2}}, \tag{2.46}$$

a result to be proved later, where the reduced matrix element, the quantity with double bars, is given by

$$(l' \| \, Y_L \, \| l) = (2l + 1)^{\frac{1}{2}}(lL00 \mid l'0), \tag{2.47}$$

The frequently occurring symbol, $(ab00 \mid c0)$, the vector addition coefficient with zero magnetic quantum numbers, has the very important property that it is zero if $(a + b + c)$ is odd.

The following symmetry properties of the vector addition coefficients are very useful:

$$\begin{aligned}
(j_1 j_2 m_1 m_2 \mid jm) &= (j_2 j_1 - m_2 - m_1 \mid j - m) \\
&= (-1)^{j_1 + j_2 - j}(j_2 j_1 m_2 m_1 \mid jm) \\
&= (-1)^{j_1 + j_2 - j}(j_1 j_2 - m_1 - m_2 \mid j - m) \\
&= (-1)^{j_1 - m_1}\left[\frac{2j + 1}{2j_2 + 1}\right]^{\frac{1}{2}}(j_1 j m_1 - m \mid j_2 - m_2) \\
&= (-1)^{j_2 + m_2}\left[\frac{2j + 1}{2j_1 + 1}\right]^{\frac{1}{2}}(j j_2 - m m_2 \mid j_1 - m_1).
\end{aligned} \tag{2.48}$$

2.2.2. Racah Coefficients

Consider a system of three particles: we can construct composite wave functions of [50]

$$\mathbf{J} = \mathbf{J}_1 + \mathbf{J}_2 + \mathbf{J}_3.$$

Such composite wave functions will not be unique but will depend on the coupling scheme. For example, if we were to couple $|j_1 m_1\rangle$ and $|j_2 m_2\rangle$ together,

$$|j_1 m_1\rangle \, |j_2 m_2\rangle = \sum_{j_{12} m_{12}} (j_1 j_2 m_1 m_2 \mid j_{12} m_{12}) \, |j_1 j_2 j_{12} m_{12}\rangle$$

and then couple the resultant to $|j_3m_3\rangle$,

$$|j_3m_3\rangle\,|(j_1j_2)j_{12}m_{12}\rangle = \sum_{jm}(j_{12}j_3m_{12}m_3\,|\,jm)\,|(j_1j_2)j_{12}j_3jm\rangle. \tag{2.49}$$

The state vector $|j_{12}j_3jm\rangle$ is the simultaneous eigenfunction of the operators $\mathbf{J}_{12}$, $\mathbf{J}_3$, $\mathbf{J}^2$, and $\mathbf{J}_z$. Alternatively, we may couple j_2 and j_3 together and then couple the resultant state vector $|j_2j_3j_{23}m_{23}\rangle$ to $|j_1m_1\rangle$ to get the final composite state vector $|j_1, (j_2j_3)j_{23}; jm\rangle$. Using the orthonormality properties of the vector addition coefficients, Eq. 2.49 can be rewritten as

$$|(j_1j_2)j_{12}j_3jm\rangle = \sum_{m_{12},m_3}(j_{12}j_3m_{12}m_3\,|\,jm)\sum_{m_1m_2}(j_1j_2m_1m_2\,|\,j_{12}m_{12})$$
$$\times\,|j_1m_1\rangle\,|j_2m_2\rangle\,|j_3m_3\rangle. \tag{2.50}$$

and similarly for $|j_1, (j_2j_3)j_{23}; jm\rangle$.

If we replace $|j_1m_1\rangle\,|j_2m_2\rangle\,|j_3m_3\rangle$ in Eq. 2.50 with the analogue to (2.49) in terms of $|j_1, (j_2j_3)j_{23}; jm\rangle$, then Eq. 2.50 becomes

$$|(j_1j_2)j_{12}, j_3; jm\rangle = \sum_{m_{12}m_3}\sum_{m_1m_2}\sum_{j_{12}m_{23}}\sum_{j'm'}(j_{12}j_3m_{12}m_3\,|\,jm)(j_1j_2m_1m_2\,|\,j_{12}m_{12})$$
$$\times\,(j_2j_3m_2m_3\,|\,j_{23}m_{23})(j_1j_{23}m_1m_{23}\,|\,j'm')\,|j_1, (j_2j_3)j_{23}; j'm'\rangle. \tag{2.51}$$

The only term in the $(j'm')$ sums which contribute are $j' = j$, $m' = m$. Consequently, Eq. 2.51 is an identity relating the wave functions of one coupling scheme to the wave functions of a different coupling scheme. We can introduce a new notation for the expansion coefficient relating one basis to the other

$$|(j_1j_2)j_{12}, j_3; jm\rangle = \sum_{j_{23}}\langle j_1, (j_2j_3)j_{23}; j\,|\,(j_1j_2)j_{12}, j_3; j\rangle\,|j_1, (j_2j_3)j_{23}; jm\rangle. \tag{2.52}$$

The expansion coefficient is also known as a recoupling coefficient and according to Eq. 2.51 it can be defined in terms of vector addition coefficients to be

$$\langle j_1, (j_2j_3)j_{23}; j\,|\,(j_1j_2)j_{12}, j_3; j\rangle = \sum_{m_1m_2m_3}(j_1j_2m_1m_2\,|\,j_{12}m_{12})(j_{12}j_3m_{12}m_3\,|\,jm)$$
$$\times\,(j_2j_3m_2m_3\,|\,j_{23}m_{23})(j_1j_{23}m_1m_{23}\,|\,jm)$$
$$\equiv [(j_{23} + 1)(2j_{12} + 1)]^{1/2}W(j_1j_2jj_3; j_{12}j_{23}), \tag{2.53}$$

where W is the Racah [51] or $6j$ coefficient, since it has six angular momentum vectors in its argument list. It is a real function.

In the Racah coefficient, the six arguments are not all independent and must satisfy the triangular inequalities for the four triads

$$(j_1 j_2 j_{12}), \quad (j_{12} j_3 j), \quad (j_2 j_3 j_{23}), \quad (j_{23} j_1 j).$$

The suffix notation makes this requirement look self-evident. By using the symmetry properties of the vector addition coefficients it can be shown that the W-coefficients have the properties

$$\begin{aligned} W(abcd; ef) &= W(badc; ef) \\ &= W(cdab; ef) \\ &= W(acbd; fe) \\ &= (-1)^{a+d-f-e} W(ebcf; ad) \\ &= (-1)^{b+c-f-e} W(aefd; bc). \end{aligned} \tag{2.54}$$

When we compute the overlap integral

$$\langle j_{12}' j_3; jm \,|\, j_{12} j_3; jm] = \sum_{j_{23}'} \sum_{j_{23}} [(2j_{23}' + 1)(2j_{12}' + 1)(2j_{23} + 1)(2j_{12} + 1)]^{1/2} \times W(j_1 j_2 j j_3; j_{12}' j_{23}') W(j_1 j_2 j j_3; j_{12} j_{23}) \langle j_1 j_{23}' jm \,|\, j_1 j_{23} jm \rangle$$

and use the fact that the state vectors are orthonormal, that is, we shall have $\delta_{j_{12}' j_{12}}$ and $\delta_{j_{23}' j_{23}}$ on the left-hand sides and right-hand sides respectively, we arrive at one of the orthonormality relations for the Racah coefficient

$$\sum_{j_{23}} (2j_{23} + 1)(2j_{12} + 1) W(j_1 j_2 j j_3; j_{12}' j_{23}) W(j_1 j_2 j j_3; j_{12} j_{23}) = \delta_{j_{12} j_{12}'}. \tag{2.55}$$

When we write out the state vectors in Eq. 2.52 in full we obtain

$$\begin{aligned} &\sum_{m_{12}\cdots} (j_{12} j_3 m_{12} m_3 \,|\, jm)(j_1 j_2 m_1 m_2 \,|\, j_{12} m_{12}) \,|j_1 m_1\rangle \,|j_2 m_2\rangle \,|j_3 m_3\rangle \\ &\qquad = \sum_{j_{23}} \langle j_1, (j_2 j_3) j_{23}; j \,|\, (j_1 j_2) j_{12}, j_3; j \rangle \\ &\qquad \times \sum_{m_1'\cdots} (j_1 j_{23} m_1' m_{23} \,|\, jm)(j_2 j_3 m_2' m_3' \,|\, j_{23} m_{23}) \,|j_1 m_1'\rangle \,|j_2 m_2'\rangle \,|j_3 m_3'\rangle \end{aligned}$$

When the orthonormality of $|j_i m_i\rangle$ is used, we obtain the very important relation

$$\begin{aligned} (j_1 j_2 m_1 m_2 \,|\, j_{12} m_{12})(j_{12} j_3 m_{12} m_3 \,|\, jm) &= \sum_{j_{23}} \langle j_1, (j_2 j_3) j_{23}; j \,|(j_1 j_2) j_{12}, j_3; j \rangle \\ &\times (j_1 j_{23} m_1 m_{23} \,|\, jm)(j_2 j_3 m_2 m_3 \,|\, j_{23} m_{23}) \end{aligned} \tag{2.56}$$

In a typical problem one is faced with summing a product of several vector addition coefficients. Successive applications of Eq. 2.56 will allow recoupling of the angular momenta until the magnetic quantum number concerned can be summed out as will now be proved. Multiplying Eq. 2.56 by $(j_2 j_3 m_2 m_3 \mid j_{23}' m_{23}')$ and summing over m_2 and m_3 results in the factor

$$\sum_{m_2 m_3} (j_2 j_3 m_2 m_3 \mid j_{23}' m_{23}')(j_2 j_3 m_2 m_3 \mid j_{23} m_{23}) = \delta_{j_{23}' j_{23}} \delta_{m_{23}' m_{23}}$$

appearing on the right-hand side, and (2.56) becomes

$$\sum_{m_2 m_3} (j_1 j_2 m_1 m_2 \mid j_{12} m_{12})(j_{12} j_3 m_{12} m_3 \mid jm)(j_2 j_3 m_2 m_3 \mid j_{23} m_{23}) = (j_1 j_{23} m_1 m_{23} \mid jm)\langle j_1, (j_2 j_3) j_{23}; j \mid (j_1 j_2) j_{12}, j_3; j\rangle \quad (2.57)$$

which will be used in Eq. 2.68 when we compute the matrix elements of the two-electron operator r_{12}^{-1}.

A useful relation between Racah coefficients can be derived by considering the recoupling relation

$$|j_1, (j_2 j_3) j_{23}; jm\rangle = \sum_{j_{12}} \langle j_1, (j_2 j_3) j_{23}; j \mid (j_1 j_2) j_{12}, j_3; j\rangle \, |(j_1 j_2) j_{12}, j_3; jm\rangle$$

and a similar result with j_1 and j_2 interchanged, namely,

$$|j_2, (j_1 j_3) j_{13}; jm\rangle = \sum_{j_{12}} \langle j_2, (j_1 j_3) j_{13}; j \mid (j_2 j_1) j_{12}, j_3; j\rangle \, |(j_2 j_1) j_{12}, j_3; jm\rangle,$$

where we have again used j_{12} rather than j_{21}, since we have a sum on the right-hand side and j_{12} and j_{21} have the same range. From the symmetry properties of the vector addition coefficients we have that

$$|(j_2 j_1) j_{12}, j_3; jm\rangle = (-1)^{j_1 + j_2 - j_{12}} |(j_1 j_2) j_{12}, j_3; jm\rangle. \quad (2.58)$$

We now compute the inner product

$$\langle j_1, (j_2 j_3) j_{23}; jm \mid j_2, (j_1 j_3) j_{13}; jm\rangle = \sum_{j_{12}, j_{12}'} \langle (j_1 j_2) j_{12}, j_3; jm \mid (j_2 j_1) j_{12}'; jm\rangle \times \langle j_1, (j_2 j_3) j_{23}; j \mid (j_1 j_2) j_{12}, j_3; j\rangle \times \langle j_2, (j_1 j_3) j_{13}; j \mid (j_2 j_1) j_{12}', j_3; j\rangle.$$

We note that the left-hand side inner product is a Racah coefficient itself, while the inner product of the right-hand side is simply $(-1)^{j_1 + j_2 - j_{12}'} \delta_{j_{12} j_{12}'}$, and we obtain

$$W(j_1 j_{23} j_{13} j_2; j j_3) = \sum_{j_{12}} (-1)^{j_1 + j_2 - j_{12}} W(j_1 j_2 j j_3; j_{12} j_{23}) W(j_2 j_1 j j_3; j_{12} j_{13}). \quad (2.59)$$

In other words, when there is a product of several Racah coefficients it may be possible to reduce the number of terms in the product.

2.2.3. Matrix Elements of r_{12}^{-1}

It is well known that the two-electron operator can be expanded in terms of Legendre polynomials

$$\frac{1}{r_{12}} = \sum_{k=0}^{\infty} \mathbf{P}_k(\hat{r}_1 \cdot \hat{r}_2)\begin{cases} \dfrac{r_1^{\,k}}{r_2^{k+1}}, & \text{when} \quad r_1 < r_2 \\ \dfrac{r_2^{\,k}}{r_1^{k+1}}, & \text{when} \quad r_1 > r_2 \end{cases} \tag{2.60}$$

We wish to calculate the matrix elements of this operator in a representation

$$|(l_1 l_2)LM\rangle = \sum_{m_1 m_2} (l_1 l_2 m_1 m_2 \mid LM)\,|l_1 m_1\rangle\,|l_2 m_2\rangle \tag{2.61}$$

constructed from the vector product of two spherical harmonics; hence

$$\langle (l_1 l_2)LM|\, r_{12}^{-1}\,|(l_1' l_2')L'M'\rangle$$
$$= \sum_{k=0}^{\infty} f_k(r_1, r_2)\langle (l_1 l_2)LM|\,\mathbf{P}_k(\hat{r}_1 \cdot \hat{r}_2)\,|(l_1' l_2')L'M'\rangle \tag{2.62}$$

In Eq. 1.29 we have expanded P_k in terms of spherical harmonics, which can also be written as the scalar product of two tensor operators,

$$P_k(\hat{r}_1 \cdot \hat{r}_2) = \mathbf{C}^k \cdot \mathbf{C}^{k*} = \sum_{q=-k}^{k} C_q^{\,k} C_q^{\,k*} \tag{2.63}$$

where

$$C_q^{\,k} = \left(\frac{4\pi}{2k+1}\right)^{1/2} Y_{kq}(\hat{r}_1)$$

and

$$C_q^{\,k*} = (-1)^q \left(\frac{4\pi}{2k+1}\right)^{1/2} Y_{k-q}(\hat{r}_2),$$

where Y_{kq} is normalized as in Eq. 1.28 and *not* as in Fano et al. [52]. Application of $C^k{}_q$ from the right on the single particle wave function $\langle l_1 m_1|$, see Eq. 2.42 for expanding out $|l_1 l_2 LM\rangle$, yields

$$\langle l_1 m_1|\, C_q^{\,k} = \sum_{lm} \langle l_1 m_1|\, C_q^{\,k}\,|lm\rangle\langle lm|$$
$$= \sum_{lm} (2l_1 + 1)^{-1/2}(l_1\,\|C^k\|\,l)(lkmq \mid l_1 m_1)\langle lm| \tag{2.64}$$

using the Wigner-Eckart theorem which introduces the reduced matrix element $(l_1\| C^k \|l)$. We can evaluate the reduced matrix element by calculating the full matrix element by an alternative method:

$$\langle l_1 m_1 | C_q^k | lm \rangle = \int d\hat{r}_1 Y^*_{l_1 m_1} \left(\frac{4\pi}{2k+1}\right)^{1/2} Y_{kq} Y_{lm}$$
$$= \left[\frac{4\pi}{2k+1} \cdot \frac{(2k+1)(2l+1)}{4\pi(2l_1+1)}\right]^{1/2} (lkmq \mid l_1 m_1)(lk00 \mid l_1 0). \tag{2.65}$$

When Eq. 2.65 is compared with (2.64) the result already quoted in Eq. 2.47 is obtained, namely

$$(l_1\| C^k \|l) = (2l+1)^{1/2}(lk00 \mid l_1 0)$$
$$= (-1)^k (l\| C^k \|l_1), \tag{2.66}$$

from the properties (2.48). Application of C_q^{k*} from the left on the single-particle wave function $|l_2' m_2'\rangle$ gives

$$C_q^{k*} |l_2' m_2'\rangle = \sum_{l'm'} |l'm'\rangle\langle l'm'| C_q^{k*} |l_2' m_2'\rangle$$
$$= \sum_{l'm'} |l'm'\rangle(-1)^q \langle l'm'| C_{-q}^k |l_2' m_2'\rangle$$
$$= \sum_{l'm'} |l'm'\rangle(-1)^q (l'\| C^k \|l_2')(2l'+1)^{-1/2}(l_2' k m_2' - q \mid l'm') \tag{2.67}$$

having used the Wigner-Eckart theorem again. When Eqs. 2.63, 2.64, and 2.67 are substituted into Eq. 2.62 we obtain

$$\langle (l_1 l_2)LM | \mathbf{P}_k(\hat{r}_1 \cdot \hat{r}_2) | (l_1' l_2')L'M' \rangle$$
$$= \sum_{q m_1 m_1' m_2 m_2'} (l_1 l_2 m_1 m_2 \mid LM)(l_1' l_2' m_1' m_2' \mid L'M')$$
$$\times \langle l_2 m_2 | \sum_{lm} (2l_1+1)^{-1/2}(l_1\| C^k \|l)(lkmq \mid l_1 m_1)\langle lm | |l_1' m_1'\rangle$$
$$\times \sum_{l'm'} (2l'+1)^{-1/2}(-1)^q (l'\| C^k \|l_2')(l_2' k m_2' - q \mid l'm') |l'm'\rangle,$$

where now the one-particle overlap integrals give $\delta_{l_2 l'}\delta_{m_2 m'}$ and $\delta_{l l_1'}\delta_{m m_1'}$. The sums over the magnetic quantum numbers can be performed using Eq. 2.57, namely,

$$\sum_{q m_1 m_2} (l_1' k m_1' q \mid l_1 m_1)(l_1 l_2 m_1 m_2 \mid LM)(l_2' k m_2' - q \mid l_2 m_2)(-1)^q$$
$$= \sum_{q m_1 m_2} (l_1' k m_1' q \mid l_1 m_1)(l_1 l_2 m_1 m_2 \mid LM)(l_2 k m_2 q \mid l_2' m_2')(-1)^k \left(\frac{2l_2+1}{2l_2'+1}\right)^{1/2}$$
$$= (-1)^k \left[\frac{(2l_2+1)}{(2l_2'+1)}\right]^{1/2} (l_1' l_2' m_1' m_2' \mid LM)\langle (l_1' k)l_1, l_2; L \mid l_1', (k l_2)l_2; L\rangle$$

where the last factor is called a recoupling coefficient which can be expressed in terms of the Racah $6j$ coefficient by

$$\langle (l_1'k)l_1, l_2; L \mid l_1', (kl_2)l_2'; L\rangle = [(2l_2'+1)(2l_1+1)]^{1/2} W(l_1'kLl_2; l_1l_2').$$

Finally, from the orthonormality of the vector addition coefficients,

$$\sum_{m_1'm_2'} (l_1'l_2'm_1'm_2' \mid L'M')(l_1'l_2'm_1'm_2' \mid LM) = \delta_{LL'}\,\delta_{MM'},$$

we have

$$\begin{aligned}\langle (l_1l_2)LM|\, P_k(\hat{r}_1\cdot\hat{r}_2)\,|(l_1'l_2')L'M'\rangle &= \delta_{LL'}\,\delta_{MM'}[(2l_1+1)(2l_2'+1)]^{-1/2}\\ &\quad\times (l_1\| C^k \| l_1')(l_2\| C^k \| l_2')(-1)^k\\ &\quad\times \langle (l_1'k)l_1, l_2; L \mid l_1', (kl_2)l_2'; L\rangle. \quad (2.69)\end{aligned}$$

We now note that: starting from Eq. 2.62 we went through the intermediate step of introducing magnetic quantum numbers, but ended up in Eq. 2.69 without them. Consequently, we should be able to devise a technique which avoids the algebra of the magnetic quantum numbers. Such a method has been developed by Fano *et al.* [52]. This method involves introducing a fictitious particle, called an "orbiton." The use of this fiction is suggested by the form of the recoupling coefficient in Eq. 2.69. That is to say, in this recoupling coefficient we are considering the transformation between coupling schemes of three vectors l_1', k, and l_2; in other words, the left- and right-hand sides of the recoupling coefficient resemble three-particle wave functions.

Let $|kq\rangle$ be the wave function of the orbiton, then we can rewrite Eq. 2.63 as

$$\begin{aligned}P_k(\hat{r}_1\cdot\hat{r}_2) &= \sum_{q=-k}^{k} C_q^k \sum_{q'=-k}^{k} \delta_{qq'} C_{q'}^{k*}\\ &= \sum_{q=-k}^{k} C_q^k \langle kq| \sum_{q'=-k}^{k} |kq'\rangle\, C_{q'}^{k*} \qquad (2.70)\end{aligned}$$

Instead of Eq. 2.64 when we apply the full operator, we obtain

$$\langle l_1m_1| \sum_q C_q^k \langle kq| = \sum_l (2l_1+1)^{-1/2}(l_1\| C^k \| l)\, \langle (lk)l_1m_1| \qquad (2.71)$$

having summed over m and q when introducing the product wave function. Therefore, when this full operator is applied to $\langle (l_1l_2)LM|$ we have

$$\langle (l_1l_2)LM| \sum_l C_q^k \langle kq| = \sum_l (2l_1+1)^{-1/2}(l_1\| C^k \| l)\, \langle (lk)l_1, l_2; LM| \quad (2.72)$$

Formulas analogous to the above are derived for the second operator. acting from the left:

$$\sum_{q'} |kq'\rangle\, C_{q'}^{k*}\, |l_2'm'\rangle$$
$$= \sum_{l'm'q'} (-1)^{q'}(l_2'km_2' - q' \mid l'm')\,|l'm'\rangle\,|kq'\rangle\,(l'\| C^k \| l_2')(2l'+1)^{-\frac{1}{2}}$$
$$= \sum_{l'm'q'} (l'km'q' \mid l_2'm_2')(-1)^k(2l_2'+1)^{-\frac{1}{2}}\,|l'm'\rangle\,|kq'\rangle\,(l'\| C^k \| l_2')$$
$$= \sum_{l'} (-1)^k(2l_2+1)^{-\frac{1}{2}}(l'\| C^k \| l_2')\,|(kl')l_2'm_2'\rangle$$
$$= \sum_{l'} (2l'+1)^{-\frac{1}{2}}(l_2'\| C^k \| l')\,|(kl')l_2'm_2'\rangle. \tag{2.73}$$

Consequently, when we apply this operator to the full wave function we obtain

$$\sum_{q'} |kq'\rangle C_{q'}^{k*}|(l_1'l_2')L'M'\rangle$$
$$= \sum_{l'} (-1)^k(2l_2'+1)^{-\frac{1}{2}}(l'\| C^k \| l_2')\,|\, l_1', (kl')l_2'; L'M'\rangle \tag{2.74}$$

which gives the matrix element as

$$\langle (l_1l_2(LM|\, P_k(\hat{r}_1\cdot\hat{r}_2)\,|(l_1'l_2')L'M'\rangle$$
$$= \sum_{ll'} (-1)^k[(2l_1+1)(2l_2'+1)]^{-\frac{1}{2}}(l_1\| C^k \| l)(l'\| C^k \| l_2')$$
$$\times \langle (lk)l_1, l_2; LM \mid l_1', (kl')l_2'; L'M'\rangle \tag{2.75}$$

Let us emphasize that k enters the coupling scheme on the left of l' in Eq. 2.74, but on the right of l in Eq. 2.73.

On the right-hand side of Eq. 2.75 we have, besides numerical coefficients and reduced matrix elements, integrals over products of two-wave functions. Each of these integrals vanishes unless it involves a pair of wave functions constructed with products of the same one-particle wave functions. This condition requires

$$l = l_1' \qquad \text{and} \qquad l' = l_2$$

so that a single term from each of the summations in (2.73) and (2.74) gives a nonvanishing contribution. Moreover, the whole expression also vanishes unless $(LM) = (L'M')$.

Returning to Eq. 2.62, we have

$$\langle (l_1l_2)LM|\, r_{12}^{-1}\,|(l_1'l_2')L'M'\rangle = \sum_{k=0}^{\infty} f_k(r_1, r_2)\,\delta_{LL'}\,\delta_{MM'}[(2l_1+1)(2l_2+1)]^{-\frac{1}{2}}$$
$$\times (l_1\| C^k \| l_1')(l_2'\| C^k \| l_2)$$
$$\times \langle (l_1'k)l_1, l_2; L \mid l_1', (kl_2)l_2'; L\rangle. \tag{2.76}$$

In many electron atomic systems the computation of the matrix elements of the two-electron operator $r_{N,N+1}^{-1}$ can be carried out by generalizing the above results [53]. Consider an unsymmetrized n-electron wave function with total angular momentum quantum numbers LM. We may represent it in the form

$$\Psi(l_1 \cdots l_N \cdots l_{N+1} \cdots l_n; .j.\, LM) = \sum_{m_1 \cdots m_N} C^{\cdots l_N \cdots L}_{\cdots m_N \cdots M} |l_1 m_1\rangle |l_2 m_2\rangle \cdots |l_N m_N\rangle \cdots |l_n m_n\rangle, \quad (2.77)$$

where the symbol $.j.$ represent the $(n - 2)$ quantum numbers required to specify the coupling of the n one-particle angular momenta, in addition to L and M. These numbers are not listed explicitly because the method to be developed does not depend on any specific coupling scheme. In an actual computation, one has to choose a scheme and stay with it throughout the evaluation of the matrix element concerned. The coefficients C are products of vector addition coefficients. Because of Eq. 2.71

$$\Psi^*(l_1 \cdots l_N \cdots l_{N+1} \cdots l_n; .j.\, LM) \sum_q C_q^{\,k} \langle kq| = (2l_N + 1)^{-1/2} \sum_l (l_N \| C^k \| l) \Psi^*(l_1 \cdots (lk) l_N \cdots l_{N+1} \cdots l_n; \cdots LM) \quad (2.78)$$

The result of the operation is thereby expanded in a series of eigenfunctions of $(n + 1)$-particles—the initial n, plus the orbiton!

An analogous formula is obtained for the other operator

$$\sum_{q'} |kq'\rangle C_{q'}^{\,k*} \Psi(l_1' \cdots l_N' \cdots l_{N+1}' \cdots l_n'; .j'.\, L'M') = \sum_{l'} (-1)^k (2l_{N+1}' + 1)^{-1/2} (l' \| C^k \| l_{N+1}') \times \Psi(l_1' \cdots l_N' \cdots (kl') l_{N+1}' \cdots l_n; .j'.\, L'M') \quad (2.79)$$

The desired matrix element is obtained by multiplying Eqs. 2.78 and 2.79 together and integrating over all the coordinates:

$$\langle l_1 \cdots l_N \cdots l_{N+1} \cdots l_n; \cdots LM | P_k(\hat{r}_N \cdot \hat{r}_{N+1}) | l_1' \cdots l_N' \cdots l_{N+1}', \cdots l_n'; \cdots L'M' \rangle = \sum_{ll'} [(2l_N + 1)(2l_{N+1}')]^{-1/2} (-1)^k (l_N \| C^k \| l)(l' \| C^k \| l_{N+1}') \times \langle l_1 \cdots (lk) l_N \cdots l_{N+1} \cdots l_n; L | l_1' \cdots l_N' \cdots (kl') l_{N+1} \cdots l_n; L' \rangle \quad (2.80)$$

The overlap integral is simply products of integrals over pairs of wave functions. The result is zero unless $l = l_N'$ and $l = l_{N+1}$. We also must have $L = L'$. Consequently Eq. 2.80 involves the recoupling coefficient

$$\langle l \cdots (l_N' k) l_N \cdots l_{N+1} \cdots l_n; L | l_1' \cdots l_N' \cdots (k l_{N+1}) l_{N+1}' \cdots l_n; L' \rangle$$

Particular applications of these results to the scattering of electrons by any atomic system are found in Smith and Morgan [54]. For example, when the nth particle, the projectile, is one of the interacting particles, with label $(N + 1)$ and the other interacting particle is found in the ρ_i subshell on the lefthand side of the matrix element and in the ρ_j subshell on the right-hand side, then for $\rho_i \leq \rho_j$ we have

$$\begin{aligned}\langle \bar{L}_1 \cdots (\bar{L}_{\rho_i} l_{\rho_i}) L_{\rho_i} \cdots \bar{L}_{\rho_j} \cdots l_i; L_i | \, P_k(\hat{r}_N \cdot \hat{r}_{N+1}) \\ \times | \bar{L}_1 \cdots \bar{L}_{\rho_i} \cdots (\bar{L}_{\rho_j} l_{\rho_j}) L_{\rho_j} \cdots l_j; L_j \rangle \\ = [(2l_{\rho_i} + 1)(2l_j + 1)]^{-1/2} (-1)^k (l_{\rho_i} \| \, C^k \, \| l_{\rho_j})(l_i \| \, C^k \, \| l_j) \\ \times \langle \bar{L}_1 \cdots (\bar{L}_{\rho_i}, (l_{\rho_j} k) l_{\rho_i}) L_{\rho_i} \cdots \bar{L}_{\rho_j} \cdots l_i; L_i \, | \\ \bar{L}_1 \cdots \bar{L}_{\rho_i} (\bar{L}_{\rho_j} l_{\rho_j}) L_{\rho_j} \cdots (k l_i) l_j; L_j \rangle \end{aligned} \tag{2.81}$$

where the bar over quantities indicates that they are quantum numbers of spectator electrons, those electrons with labels other than N and $N + 1$. Here we have used an explicit coupling scheme. We have coupled all the electrons in a particular subshell together and then coupled the subshells together from the innermost to the outer. This matrix element is found in the so-called direct potentials, V_{ij}, where i and j run over the channel indices.

When $\rho_i > \rho_j$, since the matrix element is a real number, we have

$$\begin{aligned}\langle \bar{L}_1 \cdots (\bar{L}_{\rho_j} l_{\rho_j}) L_{\rho_j} \cdots \bar{L}_{\rho_i} \cdots l_j; L_j | \, P_k(\hat{r}_N \cdot \hat{r}_{N+1}) \\ \times | \bar{L}_1 \cdots \bar{L}_{\rho_j} \cdots (\bar{L}_{\rho_i} l_{\rho_i}) L_{\rho_i} \cdots l_i; L_i \rangle \\ = [(2l_{\rho_j} + 1)(2l_i + 1)]^{-1/2} (-1)^k (l_{\rho_j} \| \, C^k \, \| l_{\rho_i})(l_j \| \, C^k \, \| l_i) \\ \times \langle \bar{L}_1 \cdots (L_{\rho_j}, (l_{\rho_i} k) l_{\rho_j}) L_{\rho_j} \cdots L_{\rho_i} \cdots l_j; L_j \, | \\ \bar{L}_1 \cdots \bar{L}_{\rho_j} \cdots (\bar{L}_{\rho_i} l_{\rho_i}) L_{\rho_i} \cdots (k l_j) l_i; L_i \rangle \end{aligned} \tag{2.82}$$

Let λ_1, λ_2, λ_3, and λ_4 denote the four subshells containing interacting electrons in the wave functions of the matrix element, with λ_1 innermost. The computation of the recoupling coefficients in Eqs. 2.81 and 2.82 involves recoupling the value l_{λ_1} of the innermost interacting subshell through to λ_2, where l_{λ_2} is "deposited," and continued recoupling of k through to $\lambda_3 = \lambda_4$.

For $\rho_i \leq \rho_j$ the exchange terms involve the evaluation of

$$\begin{aligned}\langle \bar{L}_1 \cdots (\bar{L}_{\rho_i} l_{\rho_i}(N)) L_{\rho_i} \cdots \bar{L}_{\rho_j} \cdots l_i(N+1), L_i | \, P_k(\hat{r}_N \cdot \hat{r}_{N+1}) \\ \times | \bar{L}_1 \cdots \bar{L}_{\rho_i} \cdots (L_{\rho_j} l_{\rho_j}(N+1)) L_{\rho_j} \cdots l_j(N); L_j \rangle\end{aligned}$$

which can also be evaluated readily by Eq. 2.80 to be

$$\begin{aligned}[(2l_{\rho_i} + 1)(2l_{\rho_j} + 1)]^{-1/2} (-1)^k (l_{\rho_i} \| \, C^k \, \| l_j)(l_i \| \, C^k \, \| l_{\rho_j}) \\ \times \langle \bar{L}_1 \cdots (\bar{L}_{\rho_j}, (l_i k) l_{\rho_j}) L_{\rho_j} \cdots \bar{L}_{\rho_i} \cdots l_j; L_j \, | \\ \bar{L}_1 \cdots \bar{L}_{\rho_j} \cdots (\bar{L}_{\rho_i}, (k l_j) l_{\rho_i}) L_{\rho_i} \cdots l_i; L_i \rangle \end{aligned} \tag{2.83}$$

With λ_i defined as above, then the computation of these exchange recoupling coefficients involves recoupling l_{λ_1} forward to λ_2, as in "direct." Then the strategy is to couple both $\bar{L}_{\lambda_2}$ and k on both sides of the recoupling coefficient to the core, leaving l_{λ_4} and l_{λ_3} outside the core on the left-hand and right-hand sides of the recoupling coefficient, respectively. Finally, l_{λ_4} and l_{λ_3} are recoupled from λ_2 to the outermost subshell.

Problem 16. The radial equations for the scattering of electrons by hydrogen atoms in the Γ-representation, neglecting electron exchange, are given by

$$\left[\frac{d^2}{dr^2} - \frac{l_2(l_2+1)}{r^2} + k_n{}^2\right] F_\nu^{LS\pi}(r) = \sum_{\nu'} V_{\nu\nu'}(r) F_{\nu'}^{LS\pi}(r),$$

where $\nu = (nl_1l_2)$. Derive the general form of $V_{\nu\nu'}$ in terms of Clebsch-Gordan coefficients, Racah coefficients, and Slater integrals [55]. Check three of the potentials with those of Problem 13.

2.2.4. 9*j* Coefficients

When there are four wave functions $|j_im_i\rangle$, the composite functions of definite j and m will involve two *intermediate* coupling vectors, for example,

$$\left.\begin{aligned} j_1 + j_2 &= j_{12} \\ j_3 + j_4 &= j_{34} \end{aligned}\right\} \quad {}_{12} + j_{34} = j,$$

which has a composite state vector represented by $|(j_1j_2)j_{12}, (j_3j_4)j_{34}; jm\rangle$. In fact fifteen different types of coupling are possible and there exist real unitary transformations between the composite wave functions defined by each coupling scheme. It is possible to write out the explicit form of these transformations in terms of six vector addition coefficients in complete analogy to Eq. 2.53. A familiar example is the passage from *LS* to *jj* coupling for two particle wave functions. The total angular momentum vector of particle "1" is

$$l_1 + s_1 = j_1$$

with eigenvalues $j_1(j_1 + 1)$, where the range of j_1 is

$$|l_1 - s_1|, |l_1 - s_1 + 1|, \ldots, l_1 + s_1,$$

while the range of eigenvalues of j_2 is given by

$$|l_2 - s_2|, |l_2 - s_2 + 1|, \ldots, l_2 + s_2.$$

The total angular momentum of the system is $\mathbf{j}_1 + \mathbf{j}_2 = \mathbf{j}$. An alternative way of coupling the vectors together is to compound the orbital vectors into a resultant $\mathbf{L} = \mathbf{l}_1 + \mathbf{l}_2$, and the individual spin vectors into a total spin $\mathbf{S} = \mathbf{s}_1 + \mathbf{s}_2$. We have $\mathbf{L} + \mathbf{S} = \mathbf{j}$. From all these vectors we should expect the recoupling coefficient to have the structure

$$\langle (l_1 l_2)L, (s_1 s_2)S; j \mid (l_1 s_1)j_1, (l_2 s_2)j_2; j \rangle.$$

In order to investigate this transformation between wave functions consider the transformation $\langle \mid \rangle$ defined by

$$\begin{aligned} |j_1, (j_2 j_3)j_{23}, j_{123}, j_4; jm\rangle &= \sum_{j_{12}, j_{34}} |(j_1 j_2)j_{12}, (j_3 j_4)j_{34}; jm\rangle \\ &\quad \langle j_1, (j_2 j_3)j_{23}, j_{123}, j_4; j \mid (j_1 j_2)j_{12}, (j_3 j_4)j_{34}; j\rangle \end{aligned} \tag{2.85}$$

The left-hand side can be recoupled within $(j_1, (j_2 j_3)j_{23})j_{123}$ to give

$$\begin{aligned} |j_1, (j_2 j_3)j_{23}, j_{123}, j_4; jm\rangle &= \sum_{j_{12}} |(j_1 j_2)j_{12}, j_3, j_{123}, j_4; jm\rangle \\ &\quad \langle j_1, (j_2 j_3)j_{23}, j_{123}, j_4; j \mid (j_1 j_2)j_{12}, j_3, j_{123}, j_4; j\rangle \\ &= \sum_{j_{12}, j_{33}} \langle j_1, (j_2 j_3)j_{23}, j_{123}, j_4; j \mid (j_1 j_2)j_{12}, j_3, j_{123}, j_4; j\rangle \\ &\quad \times \langle (j_1 j_2)j_{12}, j_3, j_{123}, j_4; j \mid (j_1 j_2)j_{12}, (j_3 j_4)j_{34}; j\rangle \, |j_{12} j_{34} jm\rangle \end{aligned} \tag{2.86}$$

when we perform a second recoupling keeping j_{12} together. When these last two equations are compared we have

$$\begin{aligned} &\langle j_1, (j_2 j_3)j_{23}, j_{123}, j_4; j \mid (j_1 j_2)j_{12}, (j_3 j_4)j_{34}; j\rangle \\ &\qquad = \langle j_1, (j_2 j_3)j_{23}, j_{123}, j_4; j \mid (j_1 j_2)j_{12}, j_3, j_{123}, j_4; j\rangle \\ &\qquad\quad \times \langle (j_1 j_2)j_{12}, j_3, j_{123}, j_4; j \mid (j_1 j_2)j_{12}, (j_3 j_4)j_{34}; j\rangle \\ &\qquad = \langle (j_1, (j_2 j_3)j_{23}; j_{123} \mid (j_1 j_2)j_{12}, j_3; j_{123}\rangle \\ &\qquad\quad \times \langle (j_{12} j_3)j_{123}, j_4; j \mid j_{12}, (j_3 j_4)j_{34}; j\rangle, \end{aligned} \tag{2.87}$$

where we have suppressed those couplings which are not recouplings. Each recoupling on the right-hand side of Eq. 2.87 has the structure of a Racah coefficient. The recoupling procedure which led to Eq. 2.87 is not unique. We could have recoupled from the state vector on the left-hand side of Eq. 2.87 to the vector on the right-hand side in three steps instead of two, namely:

1. Recouple j_1, j_{23}, and j_4.
2. Recouple j_2, j_3, and j_4.
3. Recouple j_1, j_2, and j_{34}.

to yield, instead of the right-hand side of Eq. 2.87,

$$\sum_{j_{234}} \langle (j_1,(j_2j_3)j_{23}; j_{123}, j_4; j \mid j_1, (j_{23}j_4)j_{234}; j\rangle \times \langle j_1, (j_{23}j_4)j_{234}; j \mid j_1, (j_2, (j_3j_4)j_{34}j_{234}; j\rangle \times \langle j_1, (j_2, (j_3j_4)j_{34})j_{234}; j \mid (j_1j_2)j_{12}, (j_3j_4)j_{34}; j\rangle. \quad (2.88)$$

By comparing Eq. 2.87 with 2.88 and dropping the redundant labels in the latter, and thereby recognize the Racah coefficient, we obtain Biedenharn's relation [56], namely,

$$W(j_1j_{12}j_{23}j_3; j_2j_{123})W(jj_{12}j_4j_3; j_{34}j_{123}) = \sum_{\lambda} (2\lambda + 1)W(j\lambda j_{12}j_2; j_1j_{34})W(j_{23}\lambda j_3j_{34}; j_4j_2)W(j\lambda j_{123}j_{23}; j_1j_4). \quad (2.89)$$

In other words, if a problem involves the product of three $6j$ coefficients summed over one of the arguments, then one must attempt to get the expression into the form of (2.89), using the symmetry properties of the Racah coefficients. It is then possible to sum out that parameter, $\lambda = j_{234}$.

Instead of the intermediate coupling scheme used to derive the result (2.89), we can follow another route which also involves three Racah coefficients with a sum over the intermediate recoupling. This expression cannot be expressed in terms of fewer Racah coefficients and so it is used to define yet another symbol—the $9j$ coefficient. In this case we go from the state $|(j_1j_2)j_{12}, (j_3j_4)j_{34}; j\rangle$ to $|(j_1j_3)j_{13}, (j_2j_4)j_{24}; j\rangle$ and in the process define the recoupling coefficient

$$\langle (j_1j_2)j_{12}, (j_3j_4)j_{34}; j \mid (j_1j_3)j_{13}, (j_2j_4)j_{24}; j\rangle = \sum_{j_{123}} \langle (j_1j_2)j_{12}, (j_3j_4)j_{34}; j \mid (j_{12}j_3)j_{123}, j_4; j\rangle \times \langle (j_{12}j_3)j_{123}, j_4; j \mid (j_1j_3)j_{13}j_2, j_{123}j_4; j\rangle \times \langle (j_1j_3)j_{13}j_2, j_{123}, j_4; j \mid (j_1j_3)j_{13}, (j_2j_4)j_{24}; j\rangle, \quad (2.90)$$

where j_{123} extends over the range from the maximum of $|j_{12} - j_3|$ and $|j_{13} - j_2|$ to the minimum of $(j_{12} + j_3)$ and $(j_{13} + j_2)$, providing the former is less than or equal to the latter.

We now introduce an alternative notation [57] for the left-hand side of the defining equation (2.90), namely,

$$\begin{bmatrix} j_1j_2j_{12} \\ j_3j_4j_{34} \\ j_{13}j_{24}j \end{bmatrix} = \sum_{j_{123}} (2j_{123} + 1)W(j_{13}jj_2j_4; j_{24}j_{123})W(j_{12}jj_3j_4; j_{34}j_{123}) \times W(j_{13}j_3j_2j_{12}; j_1j_{123}). \quad (2.91)$$

We note that when a problem reduces to a sum over a product of three or more Racah coefficients, the order of the arguments determine whether Eq. 2.91 or 2.89 is used. The symmetry properties of the $9j$ are as follows:

1. Transposition of all rows and columns leaves the coefficient unchanged; this is equivalent to reflection through the leading diagonal.
2. Interchanging a pair of rows or a pair of columns introduces a factor $(-1)^{\sigma}$, where $\sigma = j_1 + j_2 + j_3 + j_4 + j_{12} + j_{34} + j_{13} + j_{24} + j$.

When one of the elements is zero, the symmetry properties allow the $9j$ coefficient to be written as

$$\begin{bmatrix} j_1 j_2 j_{12} \\ j_3 j_4 j_{12} \\ j_{13} j_{13} 0 \end{bmatrix} = \frac{(-1)^{j_{12}+j_{13}-j_1-j_4} W(j_1 j_2 j_3 j_4; j_{12} j_{13})}{[(2j_{12}+1)(2j_{13}+1)]^{1/2}}. \tag{2.92}$$

A generalization of the Wigner $9j$ coefficient, the $12j$ coefficient, has been given by Jahn and Hope [58] and its symmetry properties have been given by Ord-Smith [59], using a Mobius strip to illustrate these properties. However, in practical calculations involving many electron systems, it is more convenient to evaluate the generalized recoupling coefficients as products of Racah coefficients as in Smith and Morgan [54] or as in Burke [60].

2.2.5. Coefficients of Fractional Parentage

In the preceding sections we have assumed quite tacitly that we can couple individual electron wave functions together one after another with the only restriction being the limits on the quantum numbers imposed by the rules of vector addition. We shall show now that this is not so when we couple equivalent electrons together. We recall that electrons in the same subshell are said to be equivalent to one another; in other words, we can have up to six $2p$-electrons, but up to ten equivalent $3d$-electrons.

Let us consider coupling two equivalent electrons together, using the usual vector addition formula described above. We obtain overall wave functions which are either antisymmetric or symmetric under the interchange of all coordinates of the two electrons according to whether $(S + L)$ is even or odd, respectively. Since a pair of electrons can only have $S = 0$ or 1, then the only values of $S + L$ are L or $L + 1$.

When the overall two-particle wave function is written as a spin-spatial product, namely

$$\Psi = \psi_{\text{space}} \times \chi_{\text{spin}}$$

we can quickly see that since Ψ must be antisymmetric under the interchange of all the coordinates of a pair of electrons, the only two possibilities are

$$\Psi_A = \begin{cases} \psi_S \chi_A, & \text{even } L,\ S = 0,\ (S + L) \text{ even} \\ \psi_A \chi_S, & \text{odd } L,\ S = 1, (S + L) \text{ even} \end{cases}$$

where the suffix $S(A)$ denotes symmetry (antisymmetry).

We conclude that the eigenfunctions of the states with $(S + L)$ even are the normalized eigenfunctions of the allowed states of l^2. (The notation to represent q electrons in the subshell of orbital angular momentum l, is l^q.) In other words, those states with $(S + L)$ odd, which could be constructed by the rules of vector addition, are excluded when we impose the Pauli principle. For example, in the carbon atom with $2p^2$, the vector coupling gives both singlet and triplet S, P, D states; but the Pauli principle reduces these six combinations to the three allowed terms ^3P, ^1D, ^1S.

If we vector couple a third individual particle wave function to the allowed states of l^2, then the eigenfunctions so obtained are in general antisymmetric only with respect to interchange of all the coordinates of the first two electrons, but not with respect to an interchange which involves the third electron. In a slight modification to the notation developed in the previous subsections, such three-particle wave functions will be written as

$$|l^2(S'L'), l; SL\rangle,$$

where $S'L'$ are allowed values only. In this state vector, l is short-hand for orbital and spin. We shall follow the arguments of Racah [61] and consider the transformation from a coupling scheme where particles 1 and 2 are coupled together to another coupling scheme where 2 and 3 are coupled together (see Section 2.2.2).

$$|l^2(S'L'), l; SL\rangle = \sum_{S''L''} \langle l^2(S'L'), l; SL \,|\, l, ll(S''L''); SL\rangle \,|l, ll(S''L''); SL\rangle, \quad (2.93)$$

where we have been careful to write ll and not l^2 on the right-hand side to emphasize that, in general, the sum over S'' and L'' will include allowed and forbidden values of $S''L''$. Consequently, the three-particle wave function $|l^2(S'L'), l; SL\rangle$ cannot be an eigenfunction of the three-electron operators. If we write out the recoupling coefficient in (2.93) explicitly in terms of Racah coefficients, then we shall have

$$\begin{aligned} &\langle (s_1l_1, s_2l_2)S'L', s_3l_3; SL \,|\, s_1l_1, (s_2l_2, s_3l_3)S''L''; SL\rangle \\ &\quad = [(2S' + 1)(2S'' + 1)]^{1/2} W(s_1s_2Ss_3; S'S'') \\ &\qquad \times [(2L' + 1)(2L'' + 1)]^{1/2} W(l_1l_2Ll_3; L'L''). \quad (2.94) \end{aligned}$$

Let us denote those eigenfunctions, as yet unknown, which are indeed eigenfunctions of the three-particle operators, by $|l^3\alpha SL\rangle$. We can then construct linear combinations of the functions given in Eq. 2.93 with coefficients which will guarantee that terms with odd values of $(S'' + L'')$ will not be present. If we write

$$|l^3\alpha SL\rangle = \sum_{S'L'} |l^2(S'L'), l; SL\rangle(l^2(S'L'), l; SL \,|\, \} l^3\alpha SL), \tag{2.95}$$

where the latter factor, denoted by $(\cdots|\}\cdots)$, is called the fractional parentage coefficient, then replacing the state vector by the right-hand side of Eq. 2.93, results in the following condition on the fractional parentage coefficient

$$\sum_{S'L'} \langle l^2(S'L'), l; SL \,|\, l, ll(S''L''); SL\rangle(l^2(S'L'), l; SL \,|\, \} l^3\alpha SL) = 0 \qquad (S'' + L'') \text{ odd}. \tag{2.96}$$

This condition is necessary and sufficient for the determination of the coefficient of fractional parentage, CFP. The number of independent nonvanishing solutions of Eq. 2.96 for a given SL equals the number of allowed terms of this kind in l^3. If this number is greater than one, the different terms are distinguished by the parameter α. The same method may be extended to the configurations l^n, if the fractional parentages of l^{n-1} are known, namely,

$$|l^n\alpha SL\rangle = \sum_{\alpha'S'L'} |l^{n-1}(\alpha'S'L'), l; SL\rangle\langle l^{n-1}(\alpha'S'L'), l; SL \,|\, \} l^n\alpha SL\rangle \tag{2.97}$$

For $l = 1$, that is, those equivalent electrons in the p-subshells, there are unique solutions to Eq. 2.96 and so the number α can be dropped. For $l = 2$, the d-subshells, α, called the seniority, is needed to distinguish the different solutions to (2.96). For three or more particles in $l = 3$, (f)-subshells, Racah pointed out that the allowed atomic terms of the same kind cannot be distinguished by seniority number alone and need another unspecified parameter, ν say. Ishidzu and Obi [62] determined the coefficients of fractional parentage of f^3 terms. The general theory for the direct evaluation of coefficients of fractional parentage using Young operators has been given by Jahn [63] and Englefield [64].

In order to illustrate the calculation of CFP's, let us consider *three* electrons in the $l = p$ subshell; substituting Eq. 2.94 into Eq. 2.96 yields

$$\sum_{S'L'} [(2S'+1)(2S''+1)]^{1/2} W(\tfrac{1}{2}\tfrac{1}{2}S\tfrac{1}{2}; S'S'')[(2L'+1)(2L''+1)]^{1/2} W(11L1; L'L'') \times (p^2(S'L'), p; SL \,|\, \} p^3\alpha SL) = 0, \qquad S'' + L'' = \text{odd}. \tag{2.98}$$

The allowed $S''L''$ combinations of p^2 are ^3P, ^1D, and ^1S, while the forbidden combinations are ^1P, ^3D, and ^3S, and the sum over $S'L'$ involves the former combinations: for $SL = {}^2P$ and $S''L'' = {}^3S$ we have from Eq. 2.98 that

$$\begin{aligned}
&[(2 \times 1 + 1)(2 \times 1 + 1)]^{1/2} W(\tfrac{1}{2}\tfrac{1}{2}\tfrac{1}{2}\tfrac{1}{2}; 11)[(2 \times 1 + 1)(2 \times 0 + 1)]^{1/2} \\
&\quad \times W(1111; 10)(p^2({}^3P), p; {}^2P \,|\, \}p^3\alpha^2P) \\
&\quad + [(2 \times 0 + 1)(2 \times 1 + 1)]^{1/2} W(\tfrac{1}{2}\tfrac{1}{2}\tfrac{1}{2}\tfrac{1}{2}; 01)[(2 \times 2 + 1)(2 \times 0 + 1)]^{1/2} \\
&\quad \times W(1111; 20(p^2({}^1D), p; {}^2P \,|\, \}p^3\alpha^2P) \\
&\quad + [(2 \times 1 + 1)(2 \times 0 + 1)]^{1/2} W(\tfrac{1}{2}\tfrac{1}{2}\tfrac{1}{2}\tfrac{1}{2}; 01) \\
&\quad \times [(2 \times 0 + 1)(2 \times 0 + 1)]^{1/2} W(1111; 00)(p^2({}^1S), p; {}^2P \,|\, \}p^3\alpha^2P) = 0.
\end{aligned}$$

Substituting the values of $W(abcd; ef)$ from Rotenberg *et al.* [48] we get

$$-\tfrac{1}{6}(p^2({}^3P), p; {}^2P \,|\, \}p^3\alpha^2P) + \frac{\sqrt{5}}{6}(p^2({}^1D), p; {}^2P \,|\, \}p^3\alpha^2P) + \tfrac{1}{6}(p_2({}^1S), p; {}^2P \,|\, \}p^3\alpha^2P) = 0, \tag{2.99}$$

which is one equation in three unknowns. A second equation can be obtained by setting $S''L'' = {}^1P$ in Eq. 2.98 and we obtain

$$\tfrac{1}{4}(p^2({}^3P), p; {}^2P \,|\, \}p^3\alpha^2P) - \frac{\sqrt{5}}{12}(p^2({}^1D), p; {}^2P \,|\, \}p^3\alpha^2P) + \tfrac{1}{6}(p^2({}^1S), p; {}^2P \,|\, \}p^3\alpha^2P) = 0 \tag{2.100}$$

while the third and final equation is obtained by setting $S''L'' = {}^3D$:

$$\frac{\sqrt{5}}{12}(p^2({}^3P), p; {}^2P \,|\, \}p^3\alpha^2P) + \frac{1}{4\sqrt{3}}(p^2({}^1D), p; {}^2P \,|\, \}p^3\alpha^2P) + \frac{\sqrt{5}}{6}(p^2({}^1S), p; {}^2P \,|\, \}p^3\alpha^2P) = 0. \tag{2.101}$$

Equations 2.99, 2.100, 2.101 are three homogeneous equations in three unknowns; consequently the solutions for the CFP's are arbitrary to within a complex constant. In order to construct three-particle wave functions normalized to unity we shall have, from Eq. 2.95, that

$$\begin{aligned}
\langle l^3\alpha' SL \,|\, l^3\alpha SL\rangle\, \delta_{\alpha\alpha'}
&= \sum_{S'L'} (l^3\alpha' SL\{ \,|\, l^2(S'L'), l; SL) \sum_{S''L''} (l^2(S''L''), l; SL \,|\, \}l^3\alpha SL) \\
&\quad \times \langle l^2(S'L'), l; SL \,|\, l^2(S''L''), l; SL\rangle \\
&= \sum_{S'L'} (l^3\alpha' SL\{ \,|\, l^2(S'L'), l; SL)(l^2(S'L'), l; SL \,|\, \}l^3\alpha SL).
\end{aligned} \tag{2.102}$$

when we apply this result to the CFP's appearing in Eqs. 2.99–2.101, we have a fourth equation

$$\sum_{S'L'} |(p^3\alpha^2P\{\,|\,p^2(S'L'), p; {}^2P)|^2 = 1, \tag{2.103}$$

having introduced the Hermitean conjugate

$$(l^n\alpha SL\{\,|\,l^{n-1}(\alpha'S'L')lSL) = (l^{n-1}(\alpha'S'L')lSL\,|\,\}l^n\alpha SL)^* \tag{2.104}$$

The four equations (2.99, 2.100, 2.101, 2.103) determine the three CFP's up to a phase factor. To see this, let us rewrite the equations in matrix form as

$$M \cdot X = 0$$

$$|X|^2 = 1$$

where X is a column matrix whose elements are the three, possibly complex CFP's. If the vector α is a solution of this system, that is,

$$M \cdot \alpha = 0, \qquad |\alpha|^2 = 1$$

then $c\alpha$, where c is a complex scalar constant, is also a solution provided

$$|c|^2 = 1,$$

since

$$M \cdot (c\alpha) = cM\alpha = 0$$

$$|c\alpha|^2 \leq |c|^2|\alpha|^2 = |c|^2.$$

In other words, if we find any solution α, then this solution is arbitrary up to a phase, $c = \exp(i\eta)$, where η is the phase. It is a matter of convention which phase we adopt. From

$$M \cdot X = 0$$

we can equate the real and imaginary parts to zero; that is to say, we have the same system of homogeneous equations for both the real and imaginary parts of X. Consequently, the ratios of the elements must be the same for both real and imaginary parts, that is,

$$X = \begin{pmatrix} x_1 \\ x_2 \\ x_3 \end{pmatrix} = \alpha\begin{pmatrix} a \\ b \\ 1 \end{pmatrix} + i\beta\begin{pmatrix} a \\ b \\ 1 \end{pmatrix} = (\alpha + i\beta)\begin{pmatrix} a \\ b \\ 1 \end{pmatrix} = \begin{pmatrix} a \\ b \\ 1 \end{pmatrix}(\alpha^2 + \beta^2)^{1/2}e^{i\delta}$$

We recall that earlier we proved that X was arbitrary up to a complex phase factor, η. We chose η in order to make X real, that is, $\eta = (-\delta + n\pi)$!

Let us summarize the above developments: we have four equations specifying the CFP's. If we obtain real solutions to this system of equations, then we have implicitly chosen our arbitrary phase factor, up to a factor ± 1. We note therefore, that each row in the table of Problem 17 can have an *overall sign change;* we have adopted Racah's convention.

Problem 17. Show that the values of $(p^3\alpha LS \mid \}p^2(L'S'), p; LS)$ are given by

LS \ $L'S'$	^1S	^1D	^3P
^4S	0	0	1
^2P	$\sqrt{2}/3$	$-\sqrt{5}/3\sqrt{2}$	$-1/\sqrt{2}$
^2D	0	$1/\sqrt{2}$	$-1/\sqrt{2}$

The above method can be extended to the configuration l^n when the fractional parentages of l^{n-1} are known:

$$\begin{aligned}\psi(l^n\alpha SL) &= \sum_{\alpha'S'L'} (l^{n-1}(\alpha'S'L')lSL \mid \}l^n\alpha SL)\psi(l^{n-1}(\alpha'S'L')lSL) \\ &= \sum_{\alpha'S'L',\alpha''S''L''} (l^{n-1}(\alpha'S'L')lSL \mid \}l^n\alpha SL) \\ &\quad \times (l^{n-2}(\alpha''S''L'')lS'L' \mid \}l^{n-1}\alpha'S'L') \\ &\quad \times \psi(l^{n-2}(\alpha''S''L'')l(S'L')lSL).\end{aligned} \tag{2.105}$$

We now recouple the "outer" pair of electrons in ψ

$$\begin{aligned}&\psi(l^{n-2}(\alpha''S''L'')l(S'L')lSL) \\ &\quad = \sum_{\bar{S}\bar{L}} \langle l^{n-2}(\alpha''S''L'')l(S'L')l;\, SL \mid l^{n-2}(\alpha''S''L''),\, ll(\bar{S}\bar{L});\, \bar{S}\bar{L}\rangle \\ &\qquad\qquad \times \psi(l^{n-2}(\alpha''S''L''),\, ll(\bar{S}\bar{L}),\, SL)\end{aligned}$$

and substitute the right-hand side into Eq. 2.105 to derive the condition

$$\begin{aligned}&\sum_{\alpha'S'L'} (l^{n-1}(\alpha'S'L')l;\, SL \mid \}l^n\alpha SL)(l^{n-2}(\alpha''S''L'')l,\, S'L' \mid \}l^{n-1}\alpha'S'L') \\ &\quad \times \langle l^{n-2}(\alpha''S''L'')l(S'L')l;\, SL \mid l^{n-2}(\alpha''S''L''),\, ll(\bar{S}\bar{L});\, SL\rangle = 0 \quad (\bar{S} + \bar{L}) \text{ odd},\end{aligned} \tag{2.106}$$

Problem 18. Show that the transformation coefficient given in Eq. 2.106 can be expressed in terms of the pair of Racah coefficients

$$[(2L' + 1)(2\bar{L} + 1)(2S' + 1)(2\bar{S} + 1)]^{1/2} W(L''lLl;\, L'\bar{L})W(S''\tfrac{1}{2}S\tfrac{1}{2};\, S'\bar{S}).$$

As discussed earlier, the system of homogeneous equations, (2.106), for the unknowns $(l^{n-1}(\alpha' S'L')l; SL\,|\,\}l^n\alpha SL)$ do not fix the phases of the eigenfunctions of the different terms. As an example of the application of Eq. 2.106 let us consider $l = 1$, $n = 4$, and set $\bar{S} = 0$ and $\bar{L} = 1$ (i.e., the nonallowed ${}^1P$ term). An allowed term of p^4 and p^2 is ${}^1D$, that is, take $(S = 0, L = 2)$ and $(\dot{S}'' = 0, L'' = 2)$, while the allowed terms of $L'S'$ are ${}^4S$, ${}^2D$, and ${}^2P$. Using these values in Eq. 2.106 yields the equation

$$(p^{4\,1}D\,|\,\}p^{3\,2}D) \times \frac{-1}{\sqrt{2}} \times (5 \times 3 \times 2 \times 1)^{1/2} W(2121; 21) W(0\tfrac{1}{2}0\tfrac{1}{2}; \tfrac{1}{2}0)$$

$$+\, (p^{4\,1}D\,|\,\}p^{3\,2}P) \times \frac{-\sqrt{5}}{3\sqrt{2}} \times (3 \times 3 \times 2 \times 1)^{1/2} W(2121; 11) W(0\tfrac{1}{2}0\tfrac{1}{2}; \tfrac{1}{2}0) = 0$$

or

$$\frac{(p^{4\,1}D\,|\,\}p^{3\,2}D)}{(p^{3\,1}D\,|\,\}p^{3\,2}P)} = (3)^{1/2} \tag{2.107}$$

When this is taken together with the normalization condition we obtain

p^4 \ p^3	${}^4S$	${}^2D$	${}^2P$
${}^1D$	0	$\sqrt{\frac{3}{2}}$	$\frac{1}{2}$

Problem 19. Calculate the remaining CFP's for the p^4 subshell and show them to be

p^4 \ p^3	${}^4S$	${}^2D$	${}^2P$
${}^3P$	$1/\sqrt{3}$	$\frac{-\sqrt{5}}{2\sqrt{3}}$	$\frac{1}{2}$
${}^1S$	0	0	+1

2.3 CROSS-SECTION FORMULAE

2.3.1. The Scattering Amplitude

Formulas for the angular distribution and total cross sections in collisions between pairs of particles have been given by Blatt and Biedenharn [32]

when L and S are not conserved separately. Such formulas are applicable to atomic collision processes when $L-S$ coupling is included. The analogous formulas for the scattering of electrons by hydrogen atoms, when both L and S are conserved, have been derived by Percival and Seaton [55]. We derive below the scattering amplitude for electrons scattered by the atomic system, when L and S are conserved. We begin by writing down the incident wave for an electron on a neutral system

$$\Psi_{\text{inc}} \sim e^{ik_n Z_p}\chi_{\frac{1}{2}m}\psi(L_1S_1M_{L_1}M_{S_1}; \mathbf{X})$$
$$\sim \sum_{SM_SLM_Ll_2} (L_1l_2M_{L_1}0 \mid LM_L)(S_1\tfrac{1}{2}M_{S_1}m \mid SM_S)\psi(\Gamma; \mathbf{X}\hat{x}_p)$$
$$\times [4\pi(2l_2+1)]^{\frac{1}{2}} i^{l_2} j_{l_2}(k_n r_p), \tag{2.108}$$

having expanded the plane wave in terms of spherical Bessel functions, j_l. From Eqs. 2.11 and 2.14 we have the full wave function for the system given by

$$\Psi \sim \sum_{\Gamma'} \psi(\Gamma'; \mathbf{X}\hat{x}_p)\frac{1}{r_p v_{n'}^{\frac{1}{2}}}[A_{\Gamma'}e^{-i\theta_{\Gamma'}} - B_{\Gamma'}e^{i\theta_{\Gamma'}}]. \tag{2.109}$$

That part of the wave function which is due to the occurrence of a reaction with an *initial* wave function $\Psi_{\text{inc}}(nL_1M_{L_1}S_1M_{S_1}m)$ is defined by

$$\Psi_{\text{reac}}(nL_1M_{L_1}S_1M_{S_1}m)$$
$$\equiv \Psi - \Psi_{\text{inc}} \sim \sum_{\Gamma'} \frac{1}{r_p v_{n'}^{\frac{1}{2}}}\psi(\Gamma'; \mathbf{X}\hat{x}_p)[A_{\Gamma'}e^{-i\theta_{\Gamma'}} - B_{\Gamma'}e^{i\theta_{\Gamma'}}]$$
$$- \sum_{n'L_1'S_1'} \delta_{nL_1S_1,n'L_1'S_1'} \sum_{S'M_s'L'M_L'l_2'} (L_1l_2'M_{L_1}0 \mid L'M_{L'})$$
$$\times (S_1\tfrac{1}{2}M_{S_1}m \mid S'M_{S'})\,\psi(\Gamma'; \mathbf{X}\hat{x}_p)[\pi(2l_2'+1)]^{\frac{1}{2}} i^{l_2'+1}$$
$$\times \frac{1}{k_{n'}r_p}[e^{-i\theta_{\Gamma'}} - e^{i\theta_{\Gamma'}}]. \tag{2.110}$$

In order that Ψ_{reac} represent outgoing spherical waves only, then the coefficients of the ingoing waves must vanish, that is,

$$\frac{1}{v_{n''}^{\frac{1}{2}}}A_{\Gamma''} - \delta_{nL_1S_1,n''L_1''S_1''}\frac{i^{l_2''+1}}{k_{n''}}[\pi(2l_2''+1)]^{\frac{1}{2}}$$
$$\times (L_1l_2''M_{L_1}0 \mid L''M_{L''})(S_1\tfrac{1}{2}M_{S_1}m \mid S''M_{S''}) = 0.$$

From the definition of the S-matrix, Eq. 2.15, we have (since $\mathbf{S}$ is diagonal in the conserved quantum numbers)

$$B_{\Gamma'} = B_{n'L_1'S_1'l_2'L'M_L'S'M_S'} = \sum_{\Gamma''} S_{\Gamma'\Gamma''}A_{\Gamma''}$$

$$= \sum_{n''L_1''S_1''l_2''} S^{L'S'\Pi'}_{n'L_1'S_1'l_2';n''L_1''S_1''l_2''}A_{n''L_1''S_1''l_2''L'M_L'S'M_S'}$$

The channel index β is introduced to represent the collection of quantum numbers $nL_1S_1l_2$, namely,

$$B_{\beta'} = \sum_{\beta''} S_{\beta'\beta''}\frac{v_{n''}^{1/2}}{k_{n''}}\, i^{l_2''+1}[\pi(2l_2''+1)]^{1/2}(L_1l_2''M_{L_1}0 \mid L'M_{L'})$$

$$\times\,(S_1\tfrac{1}{2}M_{S_1}m \mid S'M_{S'})\,\delta_{nL_1S_1,n''L_1''S_1''}$$

$$= \sum_{l_2''=|L'-L_1|}^{L'+L_1} S^{L'S'\Pi'}_{n'L_1'S_1'l_2';nL_1S_1l_2}\frac{v_n^{1/2}i^{l_2''+1}}{k_n}[\pi(2l_2''+1)]^{1/2}$$

$$\times\,(L_1l_2''M_{L_1}0 \mid L'M_{L'})(S_1\tfrac{1}{2}M_{S_1}m \mid S'M_{S'}). \tag{2.111}$$

When this result is substituted into Eq. 2.110 we have (dropping the double prime on l_2 in the first term)

$$\Psi_{\text{reac}}(nL_1M_{L_1}S_1M_{S_1}m) \sim \sum_{\Gamma'}\frac{-1}{r_pv_{n'}^{1/2}}\psi(\Gamma';\mathbf{X}\hat{x}_p)\sum_{l_2}S^{L'S'\Pi'}_{\beta'\beta}$$

$$\times\,\frac{v_n^{1/2}}{k_n}\, i^{l_2+1}[\pi(2l_2+1)]^{1/2}(L_1l_2M_{L_1}0 \mid L'M_{L'})(S_1\tfrac{1}{2}M_{S_1}m \mid S'M_{s'})e^{i\theta_{\Gamma'}}$$

$$+\sum_{\beta'}\delta_{\beta,\beta'}\sum_{SM_SLM_Ll_2}(L_1l_2M_{L_1}0 \mid LM_L)(S_1\tfrac{1}{2}M_{S_1}m \mid SM_S)\psi(\Gamma;\mathbf{X}\hat{x}_p)$$

$$\times\,[\pi(2l_2+1)]^{1/2}i^{l_2+1}\frac{1}{k_nr_p}\,e^{i\theta_\Gamma},$$

having introduced the fictitious sum $\sum_{l_2}\delta_{l_2l_2}$, and dropped the primes on the second sum in the second term. To identify the common factors in these two terms, we change the sums over $L'M_{L'}S'M_{S'}$ to sums over LM_LSM_S in the first term to obtain

$$\Psi_{\text{reac}}(nL_1M_{L_1}S_1M_{S_1}m) \sim \sum_{LM_LSM_S}\sum_{\beta'}\left(\frac{v_n\pi}{v_{n'}}\right)^{1/2}\frac{i}{rk_{pn}}\,\psi(\beta'LM_LSM_S\Pi;\mathbf{X}\hat{x}_p]$$

$$\times\sum_{l_2}(L_1l_2M_{L_1}0 \mid LM_L)(S_1\tfrac{1}{2}M_{S_1}m \mid SM_S)(2l_2+1)^{1/2}e^{i\theta_{\beta'}}i^{l_2}(-S^{LS\Pi}_{\beta'\beta}+\delta_{\beta'\beta}).$$

For a reaction in the channel $n'L_1'S_1'$, we have the reactive part of the wave function given by

$$\Psi_{\text{reac}}(nL_1M_{L_1}S_1M_{S_1}m;\, n'L_1'S_1') \sim \left(\frac{v_n\pi}{v_{n'}}\right)^{1/2} \frac{i}{r_pk_n}$$
$$\times \sum_{LM_LSM_Sl_2'l_2} \sum_{M_{L1}'m_2'} \sum_{M_{S1}'m'} (L_1'l_2'M_{L_1'}m_2' \mid LM_L)$$
$$\times (S_1'\tfrac{1}{2}M_{S_1'}m' \mid SM_S)\psi(L_1'S_1'M_{L_1'}M_{S_1'};\, \mathbf{X})\chi_{1/2m'}Y_{l_2'm_2'}(\hat{r}_p)$$
$$\times (L_1l_2M_{L_1}0 \mid LM_L)(S_1\tfrac{1}{2}M_{S_1}M_S - M_{S_1} \mid SM_S)(2l_2+1)^{1/2}e^{i\theta_{\beta'}}i^{l_2}$$
$$\times [\delta_{\beta'\beta} - S^{LS\Pi}_{\beta'\beta}],$$

having expanded the basis function in terms of the eigenfunctions of the target atomic system. This result can be rewritten more compactly as

$$\Psi_{\text{reac}}(nL_1M_{L_1}S_1M_{S_1}m \to n'L_1'S_1') \sim i\left(\frac{v_n}{v_{n'}}\right)^{1/2} \frac{e^{ik_nr_p}}{k_nr_p}$$
$$\times \sum_{M_{L1}'M_{S1}'m'} q_{n'L_1'S_1'M_{L1}'M_{S_1}'m',\, nL_1M_{L1}S_1M_{S1}m}(\hat{r}_p)$$
$$\times \psi(L_1'S_1'M_{L_1'}M_{S_1'};\, \mathbf{X})\chi_{1/2m'}(\sigma_p) \tag{2.112}$$

where the scattering amplitude for the reaction $nL_1M_{L_1}S_1M_{S_1}m \to n'L_1'M_{L_1'}S_1M_{S_1'}m'$ is given by

$$q(n'L_1'S_1'M_{L_1'}M_{S_1'}m' \leftarrow nL_1M_{L_1}S_1M_{S_1}m;\, \hat{r}_p)$$
$$\equiv \sum_{LM_LSM_Sl_2l_2'm_2'} i^{l_2-l_2'}[\pi(2l_2+1)]^{1/2}[\delta_{\beta'\beta} - S^{LS\Pi}_{\beta'\beta}]$$
$$\times (L_1'l_2'M_{L_1'}m_2' \mid LM_L)(L_1l_2M_{L_1}0 \mid LM_L)(S_1'\tfrac{1}{2}M_{S_1'}m' \mid SM_S)$$
$$\times (S_1\tfrac{1}{2}M_{S_1}m \mid SM_S)Y_{l_2'm_2'}(\hat{r}_p) \tag{2.113}$$

When we consider the scattering of electrons by ions we must modify the foregoing formulas to take the long-range interaction into account. According to Lane and Thomas [65], the incident wave is written

$$\Psi_{\text{inc}} \sim \chi_{1/2m}\psi(L_1S_1M_{L_1}M_{S_1};\, \mathbf{X})\frac{1}{rk_n}\{\pi^{1/2}C_n(\theta)\exp[i(k_nr - \eta_n\log 2k_nr + \sigma_0)]$$
$$+ \sum_{l_2} i^{l_2}[4\pi(2l_2+1)]^{1/2}e^{i\sigma_{l_2}}f_{l_2}(r)Y_{l_20}(\hat{r})\}, \tag{2.114}$$

where the incident wave is again normalized to unit amplitude,

$$C_n(\theta) \equiv \eta_n \operatorname{cosec}^2\frac{\theta}{2}\exp\left[\frac{-2i\eta_n\log\sin\theta/2}{(4\pi)^{1/2}}\right]$$
$$\eta_n \equiv \frac{-(Z-N)}{k_n},$$

where Z is the nuclear charge, N is the number of electrons, σ_l the Coulomb phase shifts, and f_l are the regular Coulomb functions, whose asymptotic forms are given in Eq. 1.67. Following the procedure already discussed for equating to zero the coefficients of the ingoing spherical waves in the reactive part of the total wave function leads to

$$\exp\,[i(k_n r - \eta_n \log 2k_n r)] \quad \text{replacing} \quad \exp\,(ik_n r)$$

in Eq. 2.112, with the scattering amplitude defined by

$$\begin{aligned} &q(nL_1S_1M_{L_1}M_{S_1}m \rightarrow n'L_1'S_1'M_{L_1}'M_{S_1}'m';\, \hat{r}_p) \equiv -C_{n'}(\theta)\pi^{1/2} \\ &\quad \times e^{i\sigma_0}\, \delta_{nL_1S_1',n'L_1'S_1'}\, \delta_{M_{L_1}'M_{S_1}'m',M_{L_1}M_{S_1}m} \\ &\quad + i \sum_{LM_LSM_Sl_2l_2'm_2'} (L_1'l_2'M_{L_1}'m_2' \mid LM_L)(L_1l_2M_{L_1}0 \mid LM_L)(S_1\tfrac{1}{2}M_{S_1}m \mid SM_S) \\ &\quad \times (S_1'\tfrac{1}{2}M_{S_1'}m' \mid SM_S)\, i^{l_2-l_2'}[\pi(2l_2+1)]^{1/2} \exp\,[i(\sigma_{l_2}+\sigma_{l_2'})] \\ &\quad \times (\delta_{\beta'\beta} - S^{LS\Pi}_{\beta'\beta})\, Y_{l_2'm_2'}(\hat{r}). \end{aligned} \tag{2.115}$$

It is quite clear that this formula includes the neutral formula, (2.113), as a special case ($C_n = 0 = \sigma_l = \eta_n$)!

2.3.2. Differential and Total Cross Sections

In Section 1.1.3 we defined the differential cross section in terms of the scattered and incident fluxes of particles. In order to calculate the fluxes for these general collision processes we use the wave mechanical expression for current (i.e., the number of particles crossing unit area in unit time),

$$\mathbf{j} = i\hbar(\psi\nabla\psi^* - \psi^*\nabla\psi). \tag{2.116}$$

When Eq. 2.108 is used for ψ, we see that the current of incident electrons with spin m is given by

$$\mathbf{j}^m_{\text{inc}} = i\hbar(-2ik_n)\,|\chi_{1/2 m}|^2. \tag{2.117}$$

When Eq. 2.112 is used for ψ, we see that the current of electrons emerging from the scattering region, with spin m', having excited the target system to state $L_1'S_1'M_{L_1}'M_{S_1}'$, is

$$\begin{aligned} \mathbf{j}^{m'}_{\text{reac}} &= i\hbar\left(\frac{k_n}{k_{n'}} \cdot \frac{1}{k_n^2 r_p^2} \cdot -2ik_{n'}\right) |q(\hat{r})|^2 \cdot |\chi_{1/2 m'}|^2 \\ &= \frac{2\hbar}{k_n r_p^2}\,|q(\hat{r})|^2\,|\chi_{1/2 m'}|^2. \end{aligned} \tag{2.118}$$

Finally we see that the differential scattering cross section for a collision in which the emerging electron crosses an element of area $r_p^2\, d\hat{r}_p$, in the direction $\hat{r}_p$ with respect to the incident beam, having changed its spin from m to m', after exciting the target from $nL_1M_{L_1}S_1M_{S_1}$ to $n'L_1'M_{L_1}'S_1'M_{S_1}'$, is given by

$$d\sigma(nL_1M_{L_1}S_1M_{S_1}m \to n'L_1'M_{L_1}'S_1'M_{S_1}'m'; \hat{r}_p)$$

$$= \frac{2\hbar}{k_n r_p^2} |q(\hat{r}_p)|^2 |\chi_{\frac{1}{2}m'}|^2 \cdot \frac{1}{2\hbar k_n |\chi_{\frac{1}{2}m}|^2} \cdot r_p^2 \, d\hat{r}_p$$

$$= \frac{1}{k_n^2} |q(\hat{r}_p)|^2 \, d\hat{r}_p, \tag{2.119}$$

since the spin functions are normalized according to

$$\langle \chi_{\frac{1}{2}m} \mid \chi_{\frac{1}{2}m'} \rangle = \delta_{mm'}$$

The differential cross section for the process of exciting the target atom from the state nL_1S_1 to $n'L_1'S_1'$ is obtained by averaging over the initial states, $M_{L_1}M_{S_1}m$, and summing over the final states, $M_{L_1'}M_{S_1'}m'$, to give

$$d\sigma(nL_1S_1 \to n'L_1'S_1'; \hat{r})$$

$$= \sum_{M_{L_1'}M_{S_1'}m'} \frac{1}{2k_n^2(2L_1+1)(2S_1+1)} \sum_{M_{L_1}M_{S_1}m} |q(\hat{r})|^2 \, d\Omega$$

$$= \frac{\pi}{k_n^2 2(2L_1+1)(2S_1+1)} \Big[\delta_{nL_1S_1,n'L_1'S_1'} \Big\{ |C_{n'}(\theta)|^2 + 2R \Big\langle -iC_{n'}(\theta)^* e^{-i\sigma_0}$$

$$\times \sum_{LM_LSM_Sl_2l_2'm_2'M_{L_1}M_{S_1}m}$$

$$\times (L_1l_2'M_{L_1}m_2' \mid LM_L)(L_1l_2M_{L_1}0 \mid LM_L)$$

$$\times (S_1\tfrac{1}{2}M_{S_1}m \mid SM_S)(S_1\tfrac{1}{2}M_{S_1}m \mid SM_S)$$

$$\times i^{l_2-l_2'}(2l_2+1)^{\frac{1}{2}} \exp i(\sigma_{l_2}+\sigma_{l_2'})(\delta_{\beta'\beta} - S^{LS\Pi}_{\beta'\beta}) Y_{l_2'm_2'}(\hat{r}) \Big\rangle \Big\}$$

$$+ \sum_{M_{L_1'}M_{S_1'}m'} \sum_{M_{L_1}M_{S_1}m} \sum_{LM_LSM_Sl_2l_2'm_2'} i^{l_2-l_2'}(2l_2+1)^{\frac{1}{2}}(L_1l_2M_{L_1}0 \mid LM_L)$$

$$\times (L_1'l_2'M_{L_1'}m_2' \mid LM_L)(S_1'\tfrac{1}{2}M_{S_1'}m' \mid SM_S)(S_1\tfrac{1}{2}M_{S_1}m \mid SM_S)$$

$$\times \exp i(\sigma_{l_2}+\sigma_{l_2'})(\delta_{\beta'\beta} - S^{LS\Pi}_{\beta'\beta}) Y_{l_2'm_2'}(\hat{r})$$

$$\times \sum_{PM_PQM_Qp_2p_2'q_2'} (-i)^{p_2-p_2'}(2p_2+1)^{\frac{1}{2}}(L_1p_2M_{L_1}0 \mid PM_P)$$

$$\times (L_1'p_2'M_{L_1'}q_2' \mid PM_P)(S_1'\tfrac{1}{2}M_{S_1'}m' \mid QM_Q)(S_1\tfrac{1}{2}M_{S_1}m \mid QM_Q)$$

$$\times \exp -i(\sigma_{p_2}+\sigma_{p_2'})(\delta_{n'L_1'S_1'p_2',nL_1S_1p_2} - S^{PQ\Delta}_{n'L_1'S_1'p_2',nL_1S_1p_2})^*$$

$$\times Y_{p_2'q_2'}(\hat{r})^* \Big] \, d\Omega,$$

where R stands for the real part of the expression enclosed in $\langle \cdots \rangle$. Performing the various runs over magnetic quantum numbers in the above expression gives the scattering intensity to be

$$\frac{d\sigma}{d\Omega}(nL_1S_1 \to n'L_1'S_1'; \hat{r}) = \frac{\pi}{2k_n^2(2L_1+1)(2S_1+1)} \delta_{nL_1S_1,n'L_1'S_1'}$$
$$\times \left\{ |C_{n'}(\theta)|^2 + 2R\left\langle -iC_{n'}(\theta)^* \sum_{LSl_2} e^{-i\sigma_0} T_{n'L_1'S_1'l_2',nL_1S_1l_2} \right.\right.$$
$$\left.\left. \times\, Y_{l_20}(\hat{r})(2L+1)(2l_2+1)^{-1/2} \right\rangle\right\}$$
$$+ \sum_{LSl_2l_2'Pp_2p_2'} (2P+1)(2L+1)[(2l_2+1)(2l_2'+1)(2p_2'+1)]^{1/2}$$
$$\times (-1)^{L-L_1'+l_2'}[4\pi(2V+1)]^{-1/2}(l_2'p_2'00 \mid VO)(l_2p_200 \mid VO)$$
$$\times W(Ll_2'Pp_2'; L_1'V)W(l_2L_1VP; Lp_2)T^{LS\Pi}_{\beta'\beta} T^{PS\Delta *}_{n'L_1'S_1'p_2',nL_1S_1p_2} Y_{VO}(\hat{r})],$$
$$(2.120)$$

where the T-matrix is defined by

$$T^{LS\Pi}_{\beta'\beta} = \exp i(\sigma_{l_2'} + \sigma_{l_2})[\delta_{\beta'\beta} - S^{LS\Pi}_{\beta'\beta}]. \qquad (2.121)$$

Problem 20. Derive Eq. 2.120, evaluating the sums over magnetic quantum numbers explicitly.

In order to calculate the total cross sections, these expressions must be integrated over the angular variables. For inelastic processes, only the last term in Eq. 2.120 contributes, to give, in units πa_0^2,

$$\sigma(nL_1S_1 \to n'L_1'S_1') = \frac{1}{2k_n^2(2L_1+1)(2S_1+1)}$$
$$\times \sum_{LSl_2l_2'} (2L+1)(-1)^{L_1+L_1'} |T^{LS\Pi}_{\beta'\beta}|^2. \qquad (2.122)$$

2.3.3. Density Matrices and Spin Polarization

The results of any physical measurement are obtained by performing a suitable average over a statistical ensemble. This "loss" of information implied by the averaging process has its theoretical analog in the introduction of the density matrix ρ. However, calculations have usually been performed without the use of density matrices even when maximum information is *not* available. This has been done by calculating as though all the observables were available and then summing over the unobserved ones. The principal advantages of density matrices are that they avoid the introduction of unnecessary variables and they can be parameterized in terms of physical quantities which define the observable properties of the system.

Part of the technical apparatus for working with density matrices was developed previously in the section on Racah algebra; this apparatus must be extended to include the algebra of tensor operators, also developed by Racah [51], since we parameterize the density matrix in terms of irreducible spin tensor moments. This presentation of the concepts on density matrices follows that of Fano [66] and Hagedorn [67], while their application to scattering processes follows the approach developed by Simon [68] for nuclear reactions.

2.3.3.1. ***Definition of the Density Matrix.*** A *pure state* is characterized by the existence of an experiment that gives a result predictable with certainty when performed on a system in that state. Mathematically, it is possible to construct a set of Hermitian operators which have the given pure state as an eigenstate. In other words, a physical system is said to be in a definite micro-state if every observable of a complete set of commuting operators, Γ, has been measured. Such a state is said to be a pure state and corresponds to a state vector, $|\gamma\rangle$, in Hilbert space, where γ denotes all quantum numbers. It is necessary to distinguish between state vectors, on the one hand, and their *representations* on the other hand. The former are defined in an abstract sense, while the latter depend on the particular choice of basis.

When it is not convenient to identify a pure state by specifying the relevant complete experiment, or its corresponding operators, the state may be identified as a linear superposition of eigenstates of any suitable complete set of commuting operators, Δ say. Let $|\delta\rangle$ be the simultaneous eigenfunctions of Δ; in practice, δ will denote a sequence of quantum numbers. Since any state vector $|\gamma\rangle$ can be decomposed into a *coherent* mixture (that is, a linear superposition) of basis vectors, we have

$$|\gamma\rangle = \sum_{\delta} |\delta\rangle \langle \delta \mid \gamma \rangle,$$

where $\langle \delta \mid \gamma \rangle$ are the *representatives* of the ket $|\gamma\rangle$. That is, the expansion coefficients $\langle \delta \mid \gamma \rangle$ identify the pure state $|\gamma\rangle$.

The mean value of any operator, Q, when the system is in the state $|\gamma\rangle$, is defined to be

$$\begin{aligned} \langle Q \rangle^{\gamma} &= \langle \gamma | \, Q \, | \gamma \rangle \\ &= \sum_{\delta,\delta'} \langle \gamma \mid \delta \rangle \langle \delta \, | Q | \, \delta' \rangle \langle \delta' \mid \gamma \rangle. \end{aligned} \tag{2.123}$$

In this expression, the representatives of the state vectors, namely $\langle \gamma \mid \delta \rangle$ *and* $\langle \delta' \mid \gamma \rangle$*, may be written together as a pair, so that each pair is subsequently* *handled like a matrix element*. That is, write

$$\langle\delta' \mid \gamma\rangle\langle\gamma \mid \delta\rangle \equiv \langle\delta'|\, \rho^\gamma \,|\delta\rangle,$$

which is the representative of ρ^γ in the $|\delta\rangle$ basis. In an abstract Hilbert space definition we should define ρ^γ as a linear operator, since it transforms bras into bras and kets into kets. We have then in an abstract sense

$$\rho^\gamma \equiv |\gamma\rangle\langle\gamma| \tag{2.124}$$

The representative of the linear operator ρ^γ so constructed is called the *density matrix* for the pure state $|\gamma\rangle$. In terms of this matrix, the expression 2.123 for the mean value can be written

$$\begin{aligned}\langle Q\rangle^\gamma &= \sum_{\delta,\delta'} \langle\delta'|\, \rho^\gamma \,|\delta\rangle\langle\delta|\, Q \,|\delta'\rangle \\ &= \sum_{\delta'} \langle\delta'|\, \rho^\gamma Q \,|\delta'\rangle \\ &= \mathrm{Tr}\,(\rho^\gamma Q).\end{aligned} \tag{2.125}$$

In setting up Eq. 2.123 we have tacitly assumed that

$$\langle\gamma \mid \gamma\rangle = 1.$$

If we relaxed this assumption, then we should have

$$\begin{aligned}\langle Q\rangle^\gamma &= \frac{\langle\gamma|\, Q \,|\gamma\rangle}{\langle\gamma \mid \gamma\rangle} \\ &= \frac{\mathrm{Tr}\,(\rho Q)}{\sum_\delta \langle\gamma \mid \delta\rangle\langle\delta \mid \gamma\rangle} \\ &= \frac{\mathrm{Tr}\,(\rho^\gamma Q)}{\mathrm{Tr}\,(\rho^\gamma)}.\end{aligned} \tag{2.126}$$

The state vector expresses our knowledge of the system. If this knowledge is complete—that is, all the mean values of a complete commuting set of observables have been measured—that state we have called a pure state. In practice, such a situation is the exception and most times one only has partial knowledge. The state is then said to be *nonpure* or *mixed*. The following procedure has been *adopted* for calculating the probability of finding a certain experimental result when a system is in a nonpure state: firstly, one must obtain the probability for such a result for all of the pure states into which the mixed state can be decomposed; secondly, one takes the average of these results, assigning a statistical weight W^γ to each of the results. Such a procedure is *defined* to be taking an *incoherent* superposition of pure states.

Symbolically these statements are expressed by

$$\langle Q\rangle = \sum_{\gamma} W^{\gamma}\langle\gamma|\, Q\,|\gamma\rangle, \qquad \sum_{\gamma} W_{\gamma} = 1, \tag{2.127}$$

where the sum is over *all* the pure states. The concept of a nonpure state can be stated in different words: it can be *defined* to be a statistical mixture of pure states. For example, consider a partially polarized beam of spin-j particles, then Σ_{γ} in Eq. 2.127 means knowing the weights W^{γ} for *all* possible spin orientations, not just the $(2j + 1)$ orientations that one usually associates with the z-axis! It should be pointed out that the description of a nonpure state as the *incoherent* superposition of pure states is not unique. For example, there is no reason why polarized light should be described as a mixture of two linear polarizations rather than two circular polarizations.

The description of nonpure states by density matrices is also possible. When Eq. 2.123 is substituted into 2.127 we obtain

$$\begin{aligned}\langle Q\rangle &= \sum_{\gamma} W^{\gamma} \sum_{\delta,\delta'} \langle\delta \,|\, \delta\rangle\langle\delta|\, Q\,|\delta'\rangle\langle\delta' \,|\, \gamma\rangle \\ &= \sum_{\delta,\delta'} \langle\delta'|\, \rho\,|\delta\rangle\langle\delta|\, Q\,|\delta'\rangle \\ &= \mathrm{Tr}\,(\rho Q),\end{aligned} \tag{2.128}$$

if the representative of the linear operator is given now by

$$\langle\delta'|\, \rho\,|\delta\rangle = \langle\delta'|\sum_{\gamma}|\gamma\rangle\, W^{\gamma}\langle\gamma|\,|\delta\rangle$$

or abstractly, the linear operator is defined by

$$\rho = \sum_{\gamma} |\gamma\rangle\, W^{\gamma}\,\langle\gamma|\,. \tag{2.129}$$

We note that (2.129) really includes (2.124) as a special case: when the mixed state is in fact a pure state all of the W^{γ} are zero except the one for the pure state concerned which has weight unity.

From the identities (2.125) and (2.128) we see that the mean value is given by the same expression whether or not we have a pure state, consequently it seems preferable to consider ρ to be defined by (2.128) rather than the linear operator forms of (2.124) and (2.129). In other words, if the mean values are known of as many independent operators $Q^{(r)}$ as there are independent parameters in $\langle\delta'|\, \rho\,|\delta\rangle$, then the elements of the density matrix may be determined by solving a system of equations like (2.128); one equation for each $\langle Q^{r}\rangle$. However, we must first ensure that our definition $\langle Q\rangle = \mathrm{Tr}\,(\rho Q)$ is an invariant; that is, it is independent of our choice of basis.

We recall that the unit operator is defined by

$$\sum_{\varepsilon} |\varepsilon\rangle\langle\varepsilon| = 1,$$

where $|\varepsilon\rangle$ is any second basis, then

$$\begin{aligned}\langle Q\rangle = \mathrm{Tr}\,(\rho Q) &= \sum_{\delta} \langle\delta|\, \rho Q\, |\delta\rangle = \sum_{\delta\varepsilon\varepsilon'} \langle\delta \mid \varepsilon\rangle\langle\varepsilon|\, \rho Q\, |\varepsilon'\rangle\, \langle\varepsilon' \mid \delta\rangle \\ &= \sum_{\varepsilon\varepsilon'} \langle\varepsilon|\, \rho Q\, |\varepsilon'\rangle\, \langle\varepsilon' \mid \varepsilon\rangle \\ &= \sum_{\varepsilon\varepsilon'} \langle\varepsilon|\, \rho Q\, |\varepsilon'\rangle\, \delta_{\varepsilon\varepsilon'} \\ &= \sum_{\varepsilon} \langle\varepsilon|\, \rho Q\, |\varepsilon\rangle = \mathrm{Tr}\,(\rho Q).\end{aligned}$$

In summary we note that the representative of the linear operator ρ depends on *two* sets of basis vectors in Hilbert space: firstly, on that basis for which we must know the probabilities W^{γ}, and secondly, on the basis which is used for the representation. If these two bases are one and the same, then ρ is diagonal and its eigenvalues are just the probabilities W^{γ}.

The requirement that $\langle Q\rangle$ is real for Hermitian operators yields:

$$\langle Q\rangle = \langle Q\rangle^{\dagger} = \sum_{ki} Q_{ki}{}^{\dagger}\rho_{ik}{}^{\dagger} = \sum Q_{ki}\rho_{ik}{}^{\dagger} = \sum Q_{ki}\rho_{ik}$$

that is, ρ is also Hermitian. Now the number of independent parameters which are needed to specify a matrix depends on N, the number of its rows and columns. N is the number of orthogonal basis states over which the trace is taken. This number may be infinite, but is often finite when only a particular property of a system is being considered. Consequently, ρ has N^2 complex elements, but the fact that it is Hermitian restricts the number of independent parameters to N^2 real parameters.

So far, all matrix formulas have been written with two subscripts. We emphasize that *each* subscript is an abbreviated mode of writing all the eigenvalues of the complete set of commuting operators. Consequently, the representative of ρ is a multi-dimensional matrix. It is possible that we know something about the state, that is, it is only partially impure. This definite information reduces the ρ-matrix, when in a diagonal representation, with respect to the "known subscripts," to zeros except for one diagonal element which is unity. We may drop this subscript from ρ and the dimension of ρ is reduced by one. For example, if a beam of particles is known to be mono-energetic, but its spin orientation is unknown, then ρ is simply a square (two-dimensional) matrix of $2j + 1$ rows and columns. Yet a further reduction in the dimension of ρ occurs if we are *not* interested in some of the quantum numbers specifying the state; for example, if we are not interested in the azimuthal distribution of a beam of particles, then

$$\rho(\alpha jm, \alpha' j' m') = 0 \qquad \text{for} \quad m \neq m'$$

(where α denotes all the other operators which commute with $\mathbf{J}^2$ and $\mathbf{J}_z$).

2.3.3.2. ***Expansion of ρ in Terms of Hermitian Operators.*** Let us define a linear operator T which transforms every initial pure state, "in," into a corresponding final pure state, "out,"

$$|\gamma\rangle_{\text{out}} = T\,|\gamma\rangle_{\text{in}},$$

$$_{\text{out}}\langle\gamma| = {}_{\text{in}}\langle\gamma|\,T^{\dagger}.$$

In the reaction

$$a + b \to c + d$$

the system before the collision can be described by a density matrix

$$\rho_{\text{in}} = \sum_{\gamma} |\gamma\rangle_{\text{in}}\,W^{\gamma}\,{}_{\text{in}}\langle\gamma|$$

which when operated on by T and $T^{\dagger}$ becomes

$$\begin{aligned} T\rho_{\text{in}}T^{\dagger} &= \sum_{\gamma} T\,|\gamma\rangle_{\text{in}}\,W^{\gamma}\,{}_{\text{in}}\langle\gamma|\,T^{\dagger} \\ &= \sum_{\gamma} |\gamma\rangle_{\text{out}}\,W^{\gamma}\,{}_{\text{out}}\langle\gamma| \\ &\equiv \rho_{\text{out}}. \end{aligned} \tag{2.130}$$

We note that this result does not depend upon the basis chosen for the representatives of ρ.

In the spin representation

$$\chi(S_aM_a)\chi(S_bM_b)$$

then the representative of ρ_{in} will be a $(2S_a + 1)(2S_b + 1) \times (2S_a + 1)(2S_b + 1)$ Hermitian matrix which can be expanded in terms of a basis set of $[(2S_a + 1) \times (2S_b + 1)]^2$ linearly independent Hermitian matrices W_μ which satisfy the orthogonality requirement

$$\text{Tr}\,(W_\mu W_\nu) = (2S_a + 1)(2S_b + 1)\,\delta_{\mu\nu}.$$

Performing the proposed expansion of ρ_{in} gives

$$\rho_{\text{in}} = \sum_{\mu} a_\mu W_\mu, \tag{2.131}$$

where the expansion coefficients are given by

$$\begin{aligned} a_\mu &= [(2S_a + 1)(2S_b + 1)]^{-1}\,\text{Tr}\,(\rho_{\text{in}}W_\mu) \\ &= [(2S_a + 1)(2S_b + 1)]^{-1}\,\text{Tr}\,(\rho_{\text{in}})\langle W_\mu\rangle_{\text{in}} \end{aligned}$$

having used Eq. 2.126. From this result we see that if the expectation values of all the basis operators in the incoming beam are known, then ρ_{in} is completely specified.

If this expression for ρ_{in} is substituted into Eq. 2.130 and the trace is taken of the resulting equation multiplied by W_ν, then we have

$$\mathrm{Tr}\,(\rho_{out}W_\nu) = \mathrm{Tr}\left(T\sum_\mu [(2S_a+1)(2S_b+1)]^{-1}\mathrm{Tr}\,(\rho_{in})\langle W_\mu\rangle_{in}W_\mu T^\dagger W_\nu\right).$$

When Eq. 2.126 is substituted into the left-hand side, then we obtain an expression for the expectation values of the basis operators W in the outgoing beam, namely,

$$\langle W_\nu\rangle_{out}\,\mathrm{Tr}\,(\rho_{out}) = \mathrm{Tr}\,(\rho_{in})[(2S_a+1)(2S_b+1)]^{-1}\sum_\mu \langle W_\mu\rangle_{in}\,\mathrm{Tr}\,(TW_\mu T^\dagger W_\mu). \tag{2.132}$$

We can derive an expression for the ratio of the traces by taking the trace of Eq. 2.130 and use Eq. 2.131 to give

$$\frac{\mathrm{Tr}\,(\rho_{out})}{\mathrm{Tr}\,(\rho_{in})} = (2S_a+1)(2S_b+1)^{-1}\sum_\mu \langle W_\mu\rangle_{in}\,\mathrm{Tr}\,(TW_\mu T^\dagger) \equiv I(\theta\phi), \tag{2.133}$$

the scattered intensity for an incident beam whose polarization is specified by $\langle W_\mu\rangle_{in}$. Equations 2.132 and 2.133 provide a complete description of all physically measurable quantities for the general reaction $a + b \rightarrow c + d$ in terms of the transition matrix T.

The preceding results have been used by Burke and Schey [69] to describe the scattering of electrons by spin-$\frac{1}{2}$ target systems. This means that ρ will be a 4×4 matrix and they chose the 16 Hermitian bases W_μ to be the following complete set of 4×4 matrices.

$$\alpha_i \equiv \sigma_i(1) \times 1, \qquad \beta_i \equiv 1 \times \sigma_i(2), \qquad \alpha_i\beta_j$$

where $\sigma_i(1)$ are the components of the Pauli spin matrices for electron (1), 1 is the 2×2 unit matrix, and $\times$ denotes the matrix outer product. From Eq. 2.131 we have the expansion coefficients of ρ in using Eq. 2.126 to be

$$\begin{aligned}
a_\mu &= \tfrac{1}{4}\,\mathrm{Tr}\,(\rho_{in}\alpha_i) = \tfrac{1}{4}\,\mathrm{Tr}\,(\rho_{in})\langle\alpha_i\rangle_{in}, && \mu = 1, 2, 3,\\
a_\mu &= \tfrac{1}{4}\,\mathrm{Tr}\,(\rho_{in}\beta_i) = \tfrac{1}{4}\,\mathrm{Tr}\,(\rho_{in})\langle\beta_i\rangle_{in}, && \mu = 4, 5, 6,\\
a_\mu &= \tfrac{1}{4}\,\mathrm{Tr}\,(\rho_{in}\alpha_i\beta_j) = \tfrac{1}{4}\,\mathrm{Tr}\,(\rho_{in})\langle\alpha_i\beta_j\rangle_{in}, && \mu = 7, \ldots, 16.
\end{aligned}$$

Furthermore, the expectation values of the observables after the collision are also given by Eq. 2.126

$$\begin{aligned}
\langle W_\mu\rangle_{out} &= \frac{\mathrm{Tr}\,(\rho_{out}W_\mu)}{\mathrm{Tr}\,(\rho_{out})}\\
&= \frac{\mathrm{Tr}\,(T\rho_{in}T^\dagger W_\mu)}{\mathrm{Tr}\,(\rho_{out})}
\end{aligned} \tag{2.134}$$

where the representative of T is in the $|\frac{1}{2}m\frac{1}{2}m'\rangle$ spin representation. Let U be the unitary transformation connecting the two representations

$$|\tfrac{1}{2}\tfrac{1}{2}SM\rangle = U\,|\tfrac{1}{2}m\tfrac{1}{2}m'\rangle = \sum_{mm'}(\tfrac{1}{2}\tfrac{1}{2}mm' \mid SM)\,|\tfrac{1}{2}m\rangle\,|\tfrac{1}{2}m'\rangle,$$

where we have shown explicitly that U can be determined in terms of Clebsch-Gordan coefficients. Then the representative of T can be transformed as follows

$$\langle m_1m_1'|\,T\,|mm'\rangle = \langle m_1m_1'|\,U^\dagger UTU^\dagger U\,|mm'\rangle = \langle Sm_1 + m_1'|\,\bar{T}\,|Sm + m'\rangle$$

where

$$\bar{T} \equiv UTU^\dagger \qquad \text{or} \qquad T = U^\dagger\bar{T}U. \tag{2.135}$$

We know U explicitly, we now need $\bar{T}$.

At the outset we require a convenient expression for the asymptotic form of the outgoing wave. If the colliding system, which is taken to be a plane wave along the z-axis, has the initial quantum numbers α (which collectively denotes $nL_1M_{L_1}S_1$), S, and M_S, then the appropriate choice can be taken from Section 2.3.1 to be

$$\begin{aligned}
&\sum_{mM_{S_1}}(S_1\tfrac{1}{2}M_{S_1}m \mid SM_S)\psi_{\text{reac}}^{nL_1M_{L_1}S_1M_{S_1}m} \\
&\equiv \Psi_{\text{reac}}^{nL_1M_{L_1}S_1;SM_S} = \Psi_{\text{reac}}^{\alpha SM_S} \\
&= \sum_{n'L_1'S_1'M_{L_1'}}\ \sum_{S'M_{S'}}\frac{i}{k_nr_p}\left(\frac{v_n}{v_{n'}}\right)^{1/2}\phi(n'L_1'M_{L_1'})\chi(S'M_{S'})e^{ik_n'r_p} \\
&\quad\times\sum_{LM_Ll_2'l_2}[\pi(2l_2+1)]^{1/2}(L_1'l_2M_{L_1}'m_2' \mid LM_L)(L_1l_2M_{L_1}0 \mid LM_L)i^{l_2-l_2} \\
&\quad\times Y_{l_2'm_2'}(\hat{r}_p)[\delta_{\beta'\beta} - S_{\beta'\beta}]\,\delta_{SS'}\,\delta_{M_SM_S'}\,.
\end{aligned}$$

Consequently, the final wave function in the state $\alpha'S'M_S'$ has the form

$$\begin{aligned}
\Psi_{\text{reac}}(\alpha SM_S \to \alpha'S'M_S') = \left(\frac{v_n}{v_n'}\right)\frac{e^{ik_n'r_p}}{r_p}\,\phi(n'L_1'M_{L_1'}) \\
\times\,\bar{T}(\alpha SM_S, \alpha'S'M_S'; \theta\phi)\chi(S'M_S'),
\end{aligned} \tag{2.136}$$

where

$$\begin{aligned}
&\bar{T}(\alpha SM_S, \alpha'S'M_S'; \theta\phi) \\
&= \frac{i}{k_n}\sum_{LM_Ll_2'l_2\Pi} i^{l_2-l_2'}[\pi(2l_2+1)]^{1/2} \\
&\quad\times(L_1l_2M_{L_1}0 \mid LM_L)(L_1'l_2'M_{L_1}'m_2' \mid LM_L)\,\delta_{SS'}\,\delta_{M_SM_S'} \\
&\quad\times[\delta_{\beta'\beta} - S_{\beta'\beta}^{SL\Pi}]Y_{l_2'm_2'}(\theta\phi),
\end{aligned} \tag{2.137}$$

which is identical to $A_S(\theta\phi)$ of Burke and Schey. When this form for $\bar{T}$ is substituted into Eq. 2.135 and the resulting expression substituted into Eq. 2.132, we obtain the Burke and Schey results for $\langle W_\mu\rangle_{\text{out}}$.

Problem 21. Write out the explicit matrix forms for T, $T^\dagger$ and the W_μ for spin-$\frac{1}{2}$ collisions, using $A_S \equiv \bar{T}(SM_S, S'M_S')$ in the actual elements of these matrices. Derive the following expression for the depolarization ratio (a measure of the extent to which either a polarized incident electron, or a polarized bound electron, loses its polarization as a result of the collision)

$$d(\theta) \equiv \frac{\langle\alpha_i\rangle_{\text{out}}}{\langle\alpha_i\rangle_{\text{in}}} = \frac{\langle\beta_i\rangle_{\text{out}}}{\langle\beta_i\rangle_{\text{in}}} = 1 - \frac{|A_1 - A_0|^2}{3\,|A_1|^2 + |A_0|^2}$$

under the two sets of conditions (1) $\langle\alpha_i\rangle_{\text{in}} \neq 0$, $\langle\beta_i\rangle_{\text{in}} = 0 = \langle\alpha_i\beta_j\rangle_{\text{in}}$; (2) $\langle\alpha_i\rangle_{\text{in}} = 0 = \langle\alpha_i\beta_j\rangle_{\text{in}}$, $\langle\beta_i\rangle_{\text{in}} \neq 0$.

When an unpolarized beam of spin-$\frac{1}{2}$ particles is scattered from unpolarized targets (spin i), the emerging beam may be polarized, that is, the spin directions of the particles in the scattered beam are not distributed isotropically. In Eq. 2.130, W_μ operates on the $(2\cdot\frac{1}{2}+1)(2\cdot i+1)$ dimensional spin space of the initial particles a and b. For an unpolarized beam impinging on an unpolarized target, all $\langle W_\mu\rangle_{\text{in}} \equiv 0$, except for the identity $\langle 1\rangle_{\text{in}} = 1$. For such a reaction we have

$$I_1(\theta\phi) = [2(2i+1)]^{-1}\,\text{Tr}\,(TT^\dagger) \tag{2.138}$$

where the suffix 1 denotes single scattering. When we compare this result with Eq. 2.119 we see immediately that the two expressions are identical for the scattered intensity from $\alpha \to \alpha'$. In the scattered beam, again resulting from unpolarized incident electrons and unpolarized target atoms, the state of polarization is given by the expectation values of $\boldsymbol{\sigma} \times \mathbf{1}$, using Eq. 2.132, where $\boldsymbol{\sigma}$ is the Pauli spin operator for the electron, that is,

$$\langle\boldsymbol{\alpha} \times \mathbf{1}\rangle_1 = [2(2i+1)I_1(\theta\phi)]^{-1}\,\text{Tr}\,(TT^\dagger\boldsymbol{\sigma} \times \mathbf{1}) \tag{2.139}$$

using the above values for $\langle W_\mu\rangle_{\text{in}}$.

Problem 22. Show that when an unpolarized electron beam is incident on unpolarized target atomic systems there is no polarization of the scattered beam.

The important result of Problem 22 means that we must always consider some degree of polarization of either projectile or target in the initial state when working in L-S coupling. If a polarized beam of electrons is incident on unpolarized targets, then the only W_μ which will have nonzero expectation values before the collision will be $\mathbf{1} \times \mathbf{1}$ and $\boldsymbol{\sigma} \times \mathbf{1}$, hence the differential cross section is not given by Eq. 2.119 but by

$$I_2 = [2(2i+1)]^{-1}[\mathrm{Tr}\,(TT^\dagger) + \langle \boldsymbol{\sigma} \times \mathbf{1}\rangle_{\mathrm{in}} \cdot \mathrm{Tr}\,(T\boldsymbol{\sigma} \times \mathbf{1}T^\dagger)]$$

and the state of polarization of the emerging beam as obtained from Eq. 2.132 is

$$\langle \boldsymbol{\sigma} \times \mathbf{1}\rangle_2 = [2(2i+1)I_2]^{-1}[\mathrm{Tr}\,(TT^\dagger\boldsymbol{\sigma} \times \mathbf{1}) + \langle \boldsymbol{\sigma} \times \mathbf{1}\rangle_{\mathrm{in}}\, \mathrm{Tr}\,(T\boldsymbol{\sigma} \times \mathbf{1}T^\dagger\boldsymbol{\sigma} \times \mathbf{1})].$$

Problem 23. Show

$$\mathrm{Tr}\,(T\boldsymbol{\sigma} \times \mathbf{1}T^\dagger) = \mathrm{Tr}\,(TT^\dagger\boldsymbol{\sigma} \times \mathbf{1})$$

See L. Wolfenstein, *Phys. Rev.* **75,** 1664 (1949) and L. Wolfenstein and J. Ashkin, *Phys. Rev.* **85,** 947 (1952).

To evaluate these expressions we note that $\mathrm{Tr}\,(TT^\dagger)$ is already given in Eq. 2.119, which leaves to be evaluated $\mathrm{Tr}\,(T\boldsymbol{\sigma} \times \mathbf{1}T^\dagger\boldsymbol{\sigma} \times \mathbf{1})$ and

$$\mathrm{Tr}\,(TT^\dagger\boldsymbol{\sigma}\times\mathbf{1}) = \sum_{SM_SS'M_{S'}S''M_{S''}} \langle SM_S|\,T\,|S'M_{S'}\rangle\langle S'M_{S'}|\,T^\dagger\,|S''M_{S''}\rangle \times \langle S''M_{S''}|\,\boldsymbol{\sigma}\times\mathbf{1}|\,SM_S\rangle$$

where Simon and Welton [70] show that

$$\langle S''M_{S''}|\,\boldsymbol{\sigma}\times\mathbf{1}\,|SM_S\rangle = -[6(2S+1)]^{1/2}(1SM_{S''}-M_SM_S\,|\,S''M_{S''}) \times W(1\tfrac{1}{2}S''1;\tfrac{1}{2}S).$$

The matrix elements of T and $T^\dagger$ are obtained from their defining Eq. 2.137, which results in

$$\begin{aligned}
&\mathrm{Tr}\,(TT\boldsymbol{\sigma}\times\mathbf{1})\\
&= \frac{-1}{k_n^{\,2}}\sum i^{l_2-l_2'+p_2-p_2'}\pi[(2l_2+1)(2p_2+1)]^{1/2}\\
&\times (L_1l_2M_{L_1}0\,|\,LM_L)(L_1p_2M_{L_1}0\,|\,PM_P)\\
&\times (L_1'l_2'M_{L_1}'m_2'\,|\,LM_L)(L_1'p_2'M_{L_1'}q_2'\,|\,PM_P)\\
&\times \delta_{SS'}\,\delta_{M_SM_{S'}}\,\delta_{S'S''}\,\delta_{M_{S'}M_{S''}}[\delta_{\beta'\beta}-S^{SL\Pi}_{\beta'\beta}][\delta-S^{S'L\Pi}]^\dagger\\
&\times Y_{l_2'm_2'}Y_{p_2'q_2'}{}^*\,[6(2S+1)]^{1/2}(1SM_{S''}-M_SM_S\,|\,S''M_{S''})W(L\tfrac{1}{2}S''1;\tfrac{1}{2}S).
\end{aligned}$$

The sums over magnetic quantum numbers have been performed previously, see Problem 20, hence

$$\mathrm{Tr}\,(TT^{\dagger}\boldsymbol{\sigma}\times\mathbf{1})$$

$$= \frac{-\pi}{k_n^{\,2}}\sum(2P+1)(2L+1)$$

$$\times\left[\frac{6(2S+1)(2l_2+1)(2l_2'+1)(2p_2'+1)}{4\pi(2V+1)}\right]^{1/2}$$

$$\times(-1)^{L-L_1'+l_2'}(l_2'p_2'00\mid V0)(l_2p_200\mid V0)$$

$$\times W(Ll_2'Pp_2';L_1'V)W(l_2L_1VP;Lp_2)$$

$$\times(1S0M\mid SM)W(1\tfrac{1}{2}S1;\tfrac{1}{2}S)T^{LS\Pi}_{\beta'\beta}T^{PS\Delta^*}_{n'L_1'S_1'p_2',nL_1S_1p_2}Y_{V0}(\hat{r}).$$

Finally we evaluate

$$\mathrm{Tr}\,(T\boldsymbol{\sigma}\times\mathbf{1}T^{\dagger}\boldsymbol{\sigma}\times\mathbf{1})$$

$$=\frac{\pi}{k_n^{\,2}}\sum(2P+1)(2L+1)6\left[\frac{(2S+1)(2S'+1)(2l_2+1)(2l_2'+1)(2p_2'+1)}{4\pi(2V+1)}\right]^{1/2}$$

$$\times(-1)^{L-L_1'+l_2'}(l_2'p_2'00\mid V0)(l_2p_200\mid V0)W(Ll_2'Pp_2';L_1'V)W(l_2L_1VP;Lp_2)$$

$$\times(1S'M_S-M_{S'}M_{S'}\mid SM_S)(1SM_{S'}-M_SM_S\mid S'M_{S'})$$

$$\times W(1\tfrac{1}{2}S1;\tfrac{1}{2}S')W(1\tfrac{1}{2}S'1;\tfrac{1}{2}S)T^{LS\Pi}_{\beta'\beta}T^{PS'\Delta*}_{n'L_1'S_1'p_2',nL_1S_1p_2}Y_{V0}(\hat{r}).$$

When the *L-S* coupling model is abandoned and spin-dependent interactions are included then the above formulas are not applicable. Instead one must use the results given in Smith and Peshkin [71]. However, at the present time no calculations have been performed for the elements of the *T*-matrix in the presence of spin-dependent interactions.

2.3.3.3. ***Expansion of ρ in Terms of Tensor Operators.*** In previous paragraphs we considered in detail the scattering of spin-$\frac{1}{2}$ projectiles by either spin-$\frac{1}{2}$ target particles, or by unpolarized target particles of arbitrary spin. For target atomic systems whose spin is greater than $\frac{1}{2}$, we must increase the number of Hermitian matrices in the basis W_μ. An alternative expansion for ρ is based on the fact that, for the spin index alone, only $(2J+1)^2$ *real* parameters are required to specify ρ. These real parameters are given in terms of the $(2J+1)^2$ linearly independent spin tensor moments $t_k^{\,q}$ for $|K|\leq q\leq 2J$. It is especially convenient to use the tensor moments, since the low-ranking tensors correspond naturally to such physical quantities as the differential cross section and polarization. Some of the simpler tensor moments in terms of spin operators **s** are

$$t_0^{\,0} = 1, \qquad t_0^{\,1} = \frac{s_z}{[s(s+1)]^{1/2}},$$

$$t_{\pm 1}^{\,1} = \frac{\mp(s_x \pm is_y)}{[2s(s+1)]^{1/2}},$$

$$t_0^{\,2} = \frac{3s_z^{\,2} - s(s+1)}{[2s(s+1)]^{1/2}}.$$

This parameterization of ρ proceeds as follows [68].

The representative of the density matrix in Eq. 2.129 can be written in the composite spin representation of a projectile, spin i, and a target, spin I, as

$$\langle S_2M_{S_2}|\,\rho\,|S_1M_{S_1}\rangle = \langle S_2M_{S_2}|\sum_{\gamma}|\gamma\rangle\, W^{\gamma}\,\langle\gamma| \quad |S_1M_{S_1}\rangle,$$

which can be written as

$$\rho(S_1M_{S_1};\, S_2M_{S_2}) \equiv \overline{\langle S_2M_{S_2}\,|\,S_1M_{S_1}\rangle}$$

where the bar denotes the average over γ. Expanding the composite spin wave function in terms of the individual projectile and target wave functions, using Clebsch-Gordan coefficients, gives ρ in terms of the density matrices for the individual particles, namely,

$$\rho(S_1M_{S_1};\, S_2M_{S_2}) = \sum_{m_im_i'}(iIm_im_I\,|\,S_1M_{S_1})(iIm_i'm_I'\,|\,S_2M_{S_2})\rho_i(m_i, m_i')\rho_I(m_I, m_I')$$

If P_r denotes the operation of a rotation r of the coordinate system, then

$$P_r\,|SM_S\rangle = \sum_{\nu} D^S_{\gamma M_S}(r)\,|S\gamma\rangle,$$

where D is an element of the rotation group such that

$$D_{\alpha\beta}{}^S(r)D_{\gamma\delta}{}^T(r) = \sum_{\nu}(ST\alpha\gamma\,|\,\nu\alpha+\gamma)(ST\beta\delta\,|\,\nu\beta+\delta)D^{\nu}_{\alpha+\gamma,\,\beta+\delta}(r).$$

Applying P_r to the density matrix, we have, using the above relation for the product of two D's:

$$P_r\rho(S_1M_{S_1};\, S_2M_{S_2}) = \sum_{\lambda-\lambda',\,q}(-1)^{-M_{S_2}}(S_1S_2M_{S_1} - M_{S_2}\,|\,qM_{S_1} - M_{S_2}) \times D^q_{\lambda-\lambda',\,M_{S_1}-M_{S_2}}(r)t^q_{\lambda-\lambda'}$$

where

$$t^q_{\lambda-\lambda'} \equiv \sum_{\lambda'}(-1)^{\lambda'}(S_1S_2\lambda - \lambda'\,|\,q\lambda - \lambda')\overline{\langle S_2\lambda'\,|\,S_1\lambda\rangle},$$

and in the limit of zero rotation

$$\rho(S_1M_{S_1}, S_2M_{S_2}) = \sum_q (-1)^{-M_{S_2}}(S_1S_2M_{S_1} - M_{S_2} \mid qM_{S_1} - M_{S_2})t^q_{M_{S_1}-M_{S_2}}$$

which is the derived expansion for ρ. Using the unitary property of the Clebsch-Gordan coefficients, this equation can be recast giving t in terms of ρ, where ρ itself can be written out in terms of ρ_i and ρ_I, the density matrices of the individual particles taking part of the reaction. Obviously ρ_i and ρ_I can be expressed in terms of their individual tensor moments, t^i and t^I, respectively. When the Racah algebra involved is carried out [68] we obtain ρ in terms of t^i, t^I, and $9j$ coefficients.

When Eq. 2.128 is used, the measured value of any spin tensor operator $t^{q'}_{k'}$ in the *final* channel α_1' is given by

$$\begin{aligned}\langle\alpha_1'| t^{q'}_{k'} |\alpha_1'\rangle &= \mathrm{Tr}\,(\rho_{\mathrm{out}}t^{q'}_{k'}) \\ &= \sum_{S_1'M_{S_1}'} \langle\alpha_1'S_1'M_{S_1}' \mid \rho_{\mathrm{out}}t^{q'}_{k'} \mid\alpha_1'S_1'M_{S_1}'\rangle \\ &= \sum_{\alpha_2'S_1'M_{S_1}'S_2'M_{S_2}'} \langle\alpha_1'S_1'M_{S_1}'\mid \rho_{\mathrm{out}} \mid\alpha_2'S_2'M_{S_2}'\rangle \\ &\quad \times \langle\alpha_2'S_2'M_{S_2}'\mid t^{q'}_{k'} \mid\alpha_1'S_1'M_{S_1}'\rangle\end{aligned}$$

When Eq. 2.130 relating ρ_{out} to ρ_{in}, is used, we have

$$\begin{aligned}&\langle\alpha_1'S_1'M_{S_1}'\mid \rho_{\mathrm{out}} \mid\alpha_2'S_2'M_{S_2}'\rangle \\ &\quad = \langle\alpha_1'S_1'M_{S_1}'\mid T\rho_{\mathrm{in}}T^\dagger \mid\alpha_2'S_2'M_{S_2}'\rangle \\ &\quad = \sum_{\alpha_1\alpha_2S_1M_{S_1}S_2M_{S_2}} \langle\alpha_1'S_1'M_{S_1}'\mid T \mid \alpha_1S_1M_1\rangle\langle\alpha_1S_1M_{S_1}\mid \rho_{\mathrm{in}} \mid\alpha_2S_2M_{S_2}\rangle \\ &\quad\quad \times \langle\alpha_2S_2M_{S_2}\mid T^\dagger \mid\alpha_2'S_2'M_{S_2}'\rangle\end{aligned}$$

which gives

$$\begin{aligned}\langle\alpha_1'| t^{q'}_{k'} |\alpha_1'\rangle = \sum_{\alpha_1S_1'\cdots} & T(\alpha_1S_1M_{S_1}, \alpha_1'S_1'M_{S_1}'; \theta\phi)T^\dagger(\alpha_2S_2M_{S_2}, \alpha_2'S_2'M_{S_2}'; \theta\phi) \\ & \times \langle S_1M_{S_1}\mid \rho_{\mathrm{in}} \mid S_2M_{S_2}\rangle\, \delta_{\alpha_1\alpha_2}\langle S_2'M_{S_2}'\mid t^{q'}_{k'} \mid S_1'M_{S_1}'\rangle\, \delta_{\alpha_2'\alpha_1'}.\end{aligned}$$

Finally, the value of any operator $t^{q'}_{k'}$ in the final channel, when the target was initially in the state α, is

$$\begin{aligned}\langle\alpha'| t^{q'}_{k'} |\alpha'\rangle^{(\alpha)} = \sum_{S_1'\cdots} & T(\alpha S_1M_{S_1}, \alpha'S_1'M_{S_1}'; \theta\phi)T^\dagger(\alpha S_2M_{S_2}, \alpha'S_2'M_{S_2}'; \theta\phi) \\ & \times \langle S_1M_{S_1}\mid \rho_{\mathrm{in}} \mid S_2M_{S_2}\rangle\langle S_2'M_{S_2}'\mid t^{q'}_{k'} \mid S_1'M_{S_1}'\rangle,\end{aligned}$$

where the representative of ρ_{in} is given by the technique sketched above, the matrix elements of T are defined in Eq. 2.137, and the spin matrix elements of $t^{q'}_{k'}$ on the right-hand side are evaluated in ref. 70.

2.3.4. Many-Channel Photoionization

Let us examine the quantum numbers associated with the absorption of photons by atomic nitrogen in its ground state

$$\nu + N(^4S^0) \rightarrow N(^4P^e)^* \rightarrow N^+(?) + e,$$

where the question mark indicates that the residual ion may be in any of the three atomic terms $^3P^e$, $^1D^e$, or $^1S^e$ associated with the carbon-like configuration $1s^2 2s^2 2p^2$; or possibly in other terms if excited configurations of N^+ are included in the model for the final-state wave function. The quantum numbers of the intermediate state are known because absorption of a photon does not affect the spin of the system, but the parity and orbital angular momentum are changed since nonrelativistically the photon has the characteristics of a p-wave. To determine explicitly the residual atomic terms of N^+, we must take the vector composition of L^* and S^* for N^* with the allowed L_Γ and S_Γ for N^+ to yield the channel orbital angular momentum l_Γ of the ejected electron, the spin $\frac{1}{2}$ of this electron, and ensure parity is conserved in the intermediate and final states. In Table 6 we present all these quantum numbers and the reason for excluding the deleted entries when N^+ can be left in its two lowest configurations.

From this table we see that if we only include the ground state configuration of N^+, then there are only two degenerate channels $F(^3P^e, l = 0)$ and $F(^3P^e, l = 2)$. However, when we include the first excited configuration, we can have up to seven final-state channels: the above pair, together with $F(^5S^0, l = 1)$, $F(^3S^0, l = 1)$, $F(^3D^0, l = 1)$, $F(^3D^0, l = 3)$, and $F(^3P^0, l = 1)$. In other words, the photon energy is distributed between the N^+ ion and the ejected photoelectron: those photoelectrons associated with the excited $^3P^0$ term of N^+ having the least energy. Of course, at certain photon energies the excited terms of N^+ will be energetically inaccessible, but mathematically we

Table 6. Channel Quantum Numbers in the Final State Associated with Photoabsorption of Atomic Nitrogen

Target + Photon	Residual Ion, N^+	l_Γ, Photoelectron
$N(^4P^e)^*$	$1s^2 2s^2 2p^2(^3P^e)$	0, ~~1~~, 2 (parity)
	$1s^2 2s^2 2p^2(^1D^e)$	~~1~~, ~~2~~, ~~3~~ (spin)
	$1s^2 2s^2 2p^2(^1S^e)$	~~1~~ (spin)
	$1s^2 2s 2p^3(^5S^0)$	1
	$1s^2 2s 2p^3(^3S^0)$	1
	$1s^2 2s 2p^3(^3D^0)$	1, ~~2~~, 3 (parity)
	$1s^2 2s 2p^3(^1D^0)$	~~1~~, ~~2~~, ~~3~~ (spin)
	$1s^2 2s 2p^3(^3P^0)$	~~0~~, 1, ~~2~~ (parity)
	$1s^2 2s 2p^3(^1P^0)$	~~0~~, ~~1~~, ~~2~~ (spin)

must include such "virtually excited" terms—the so-called closed channels. The final-state wave function is usually generated by performing a scattering calculation, but ensuring that the resulting functions obey the correct asymptotic boundary condition; see Eq. 1.254 for the conditions for the single-channel problem.

The many-channel analog to Eq. 1.254 can be derived starting from Eq. 2.14

$$F_\Gamma(r) \sim [A_\Gamma e^{-i\theta_\Gamma} - B_\Gamma e^{i\theta_\Gamma}](2\pi k_\Gamma)^{-1/2}$$

with the S-matrix defined by

$$B_\Gamma \equiv \sum_{\Gamma'(\text{ingoing})} S_{\Gamma\Gamma'} A_{\Gamma'},$$

where we have noted explicitly that we sum over all the ingoing channels. Since S is unitary, then we have

$$A_{\Gamma'} = \sum_{\Gamma(\text{outgoing})} (S^\dagger)_{\Gamma'\Gamma} B_\Gamma$$

where the B_Γ will be specified by the asymptotic boundary condition that the reactive part of the wave function, ψ_{reac}, now represents ingoing spherical waves only, see Eq. 1.253 for the single-channel result. When the preceding form for A_Γ is substituted into Eq. 2.14 we obtain

$$F_\Gamma(r) \sim \sum_{\Gamma''} - B_{\Gamma''}[\delta_{\Gamma\Gamma''} e^{i\theta_\Gamma} - S^\dagger_{\Gamma\Gamma''} e^{-i\theta_\Gamma}](2\pi k_\Gamma)^{-1/2}$$

This result is used in Eq. 2.12 to give the final states described by the properly antisymmetrized function Ψ^f, defined by

$$\begin{aligned}
\Psi^f &\equiv (N+1)^{-1/2} \sum_p (-1)^{N+1-p} \sum_\Gamma \frac{\psi(\Gamma; \mathbf{X}\hat{x}_p) F_\Gamma(r_p)}{r_p} \\
&\sim - \sum_{\Gamma''} B_{\Gamma''} \sum_\Gamma (N+1)^{-1/2} \sum_p (-1)^{N+1-p} \psi(\Gamma; \mathbf{X}\hat{x}_p) \\
&\quad \times [\delta_{\Gamma\Gamma''} e^{i\theta_\Gamma} - S^\dagger_{\Gamma\Gamma''} e^{-i\theta_\Gamma}] \frac{(2\pi k_\Gamma)^{-1/2}}{r_p} \\
&\equiv - \sum_{\Gamma''} B_{\Gamma''} \Psi^f(\Gamma'', E, \mathbf{x}_1, \ldots, \mathbf{x}_{N+1}),
\end{aligned} \tag{2.140}$$

where k_Γ is the wave number of an electron ejected in the Γ channel and is related to the incident frequency and the Γth ionization potential I_Γ by the Einstein relation

$$\frac{(\hbar k_\Gamma)^2}{2m} = h\nu - I_\Gamma \equiv E.$$

2.3.4.1. ***Total Cross Sections.*** According to Henry and Lipsky [72], the above asymptotic form of the final-state wave function guarantees the following normalization

$$\int d\tau \Psi(\Gamma, E, \tau)^* \Psi(\Gamma', E', \tau) = \delta(E - E')\, \delta_{\Gamma\Gamma'},$$

which leads to the cross section for photoabsorption in the state $LS\Pi$ with electron ejection into the Γ channel given by (in the dipole velocity form)

$$\sigma_v(E, LS\Pi, \Gamma) = \frac{2\pi e^2 \hbar^2}{m^2 c v 2(2L+1)(2S+1)} \sum_{M_L M_S m_1}$$
$$\times \left| \left\langle \Psi(\Gamma, E, \tau) \sum_{i=1}^{N+1} \nabla_i^{m_1} \Phi(LSM_L M_S \Pi, \tau) \right\rangle \right|^2 \qquad (2.141)$$

where Φ is the target function normalized to unity, and the dipole velocity operator is given by

$$\nabla_i^{\pm 1} = \frac{1}{\sqrt{2}} \left[\mp \frac{\partial}{\partial x_i} - i \frac{\partial}{\partial y_i} \right],$$

which is replaced by

$$\left(\frac{2\pi v m}{\hbar} \right)^2 r_i Y_{1m_1}(\hat{r}_i)$$

if we require the dipole-length form of the cross section. The total cross section at energy E is obtained by summing Eq. 2.141 over $\Gamma \equiv \{n_\Gamma L_1^\Gamma S_1^\Gamma l_\Gamma L_\Gamma S_\Gamma M_{S_\Gamma} \Pi_\Gamma\}$, where $L_1^\Gamma S_1^\Gamma$ are the total quantum numbers of the residual ion and l_Γ denotes the different angular momentum states of the outgoing electron.

Since Ψ is complex, it is inconvenient for use in numerical computations. Let $\tilde{\Psi}$ be the real standing wave function for the system in the final state Γ'; this function has already been introduced in Eq. 2.12. We now interpret $F_{\Gamma\Gamma'}(r)$ as representing the radial motion of the photoelectron relative to the residual ion. Currently, there are two approaches to determining F: either Eq. 2.12 is substituted as it stands into a variational principle and the Euler equations for the F's are derived, or the F's are constrained to be orthogonal to the bound orbitals of the residual ion and another function, Φ, is added to Eq. 2.12. The function Φ is constructed wholly of the known bound orbitals of the residual ion. The consequences of the former approach have been considered by Morgan [73], while the latter approach has been adopted by Smith and Morgan [74] and is used here. Instead of Eq. 2.12 we write the antisymmetrized function in the form

$$\tilde{\Psi}(\Gamma', E, \mathbf{x}_1, \ldots, \mathbf{x}_{N+1}) = (N+1)^{-1/2} \sum_p (-1)^{N+1-p} \sum_\Gamma \frac{\psi(\Gamma; \mathbf{X}\hat{x}_p) F_{\Gamma\Gamma'}(r_p)}{r_p} + \sum_\mu C_{\Gamma'\mu} \Phi_\mu(\Gamma', \mathbf{x}_1, \ldots, \mathbf{x}_{N+1}), \quad (2.142)$$

where μ ranges over the bound configurations which can be formed by adding an electron to one of the incomplete subshells included in ψ. For example, from Table 6 we see that we would have only two such configurations, namely $1s^2 2s^2 2p^3$ and $1s^2 2s 2p^4$. The former configuration would give rise to the three atomic terms $^4S^0$, $^2D^0$, and $^2P^0$, none of which would contribute to the $\tilde{\Psi}(^4P^e)$ final state being considered in Table 6. However, the excited configuration $1s^2 2s 2p^4$ does have a $^4P^e$ atomic term and this particular Φ will be included in Eq. 2.142. The coefficients $C_{\Gamma'\mu}$ are determined from the μ-conditions

$$\sum_\nu A_{\mu\nu} C_{\Gamma_\mu'} + \sum_\Gamma \int V_{\mu,\Gamma}(r) F_{\Gamma\Gamma'}(r)\, dr = 0, \quad (2.143)$$

where both $A_{\mu\nu}$ and $V_{\mu,\Gamma}$ are known [74]. Physically speaking, we can regard the $(N+1)$ electron functions Φ as taking into account short-range correlation effects. (Of course, a nonorthogonalized F will be doing this implicitly.)

In order to relate the real function $\tilde{\Psi}$ with the complex function Ψ used in Eq. 2.141 we take the asymptotic form of Eq. 2.142, that is,

$$\tilde{\Psi}(\Gamma', E, \tau) \underset{r_{N+1}\to\infty}{\sim} (N+1)^{-1/2} \sum_\Gamma \psi(\Gamma; \mathbf{X}\hat{x}_{N+1}) \frac{[\delta_{\Gamma'\Gamma} \sin\theta_\Gamma + R_{\Gamma'\Gamma}\cos\theta_\Gamma] k_\Gamma^{-1/2}}{r_{N+1}},$$

expand the sine and cosine functions in terms of exponentials and collect terms to give

$$\tilde{\Psi}(\Gamma', E, \tau) \sim (N+1)^{-1/2} \sum_\Gamma \psi(\Gamma; \mathbf{X}\hat{x}_{N+1}) \times \frac{[(-i\delta_{\Gamma'\Gamma} + R_{\Gamma'\Gamma}) e^{i\theta_\Gamma} + (i\delta_{\Gamma'\Gamma} + R_{\Gamma'\Gamma}) e^{-i\theta_\Gamma}]}{2 r_{N+1} k_\Gamma^{1/2}}$$

We now multiply this equation by

$$\left(\frac{2}{\pi}\right)^{1/2} (-i\delta + R)^{-1}_{\Gamma''\Gamma'}$$

and sum over the open channels characterized by Γ' to obtain

$$\begin{aligned} \sum_{\Gamma'} \left(\frac{2}{\pi}\right)^{1/2} & (-i\delta + R)^{-1}_{\Gamma''\Gamma'} \tilde{\Psi}(\Gamma', E, \tau) \\ & \sim (N+1)^{-1/2} \sum_\Gamma \psi(\Gamma; \mathbf{X}\hat{x}_{N+1}) \frac{[\delta_{\Gamma''\Gamma} e^{i\theta_\Gamma} - S^\dagger_{\Gamma''\Gamma} e^{-i\theta_\Gamma}]}{(2\pi k_\Gamma)^{1/2} r_{N+1}} \\ & \equiv \Psi(\Gamma'', E, \tau), \end{aligned} \quad (2.144)$$

which can be seen from Eq. 2.140.

The total cross section for the photoabsorption of an atomic system in the state $LS\Pi$, in the dipole velocity approximation, is obtained by substituting Eq. 2.144 into Eq. 2.141 to obtain

$$\sigma_v(E, LS\Pi)\left[\frac{2\pi e^2\hbar^2}{m^2cv2(2L+1)(2S+1)}\right]^{-1}$$
$$= \frac{2}{\pi}\sum_{\Gamma}\sum_{M_LM_Sm_1}\sum_{\Gamma'\Gamma''}(i\delta + R)^{-1}_{\Gamma\Gamma'}(-i\delta + R)^{-1}_{\Gamma\Gamma''}$$
$$\times \left\langle \tilde{\Psi}(\Gamma') \sum_i \nabla_i^{m_1}\Phi(LS\Pi)\right\rangle\left\langle \tilde{\Psi}(\Gamma'') \sum_i \nabla_i^{m_1}\Phi(LS\Pi)\right\rangle^* \quad (2.145)$$

Equation 2.145 is a model-independent formula! That is to say, any theory which aspires to interpret recent experimental observations [75] must be based on a many-channel approach. The different models vary according to the degree of sophistication in computing $\tilde{\Psi}$ and Φ. The simplest model is based on the Born approximation, referred to as the Coulomb-Born approximation when the asymptotic Coulomb effect on the plane wave is taken into account. "Prescriptions" do exist for producing a symmetric **R** matrix for use with the Born approximation. The most accurate approximation available to date is the close-coupling approximation. Even within this approximation one can make the simplifying assumption that the same bound electronic orbitals are used in both Φ and $\tilde{\Psi}$. If we abandon this assumption and use the bound orbitals of the residual ion in $\tilde{\Psi}$, then we are led to the so-called "relaxed core" approximation. It is clear that this approximation neglects the effect of the ejected photoelectron on the orbitals of the residual ion. To include this effect we should have to either solve the Hartree-Fock equations for the bound orbitals, P_{nl}, at the time we solve for the channel functions F or amalgamate the close-coupling approximation with the method of polarized orbitals [76]. Within the relaxed core approximation there is still a great deal of variety in choosing $\tilde{\Psi}$. The most sophisticated calculations reported to date are based on the analysis of Henry and Lipsky [72] in which both Φ and $\tilde{\Psi}$ are taken to be single configurations. The analysis for including configuration interaction in $\tilde{\Psi}$ has been given by Smith and Morgan [74], see Eq. 2.142.

The first step in the evaluation of Eq. 2.145 involves simplifying the R-dependent factors, using the expansion

$$(1 \pm iR)^{-1}_{\Gamma\Gamma''} = (1 + R^2)^{-1}_{\Gamma\Gamma''} \mp i\left[\frac{R}{(1+R^2)}\right]_{\Gamma\Gamma''},$$

which yields

$$(1 - iR)^{-1}_{\Gamma\Gamma'}(1 + iR)^{-1}_{\Gamma\Gamma''} = (1 + R^2)^{-1}_{\Gamma\Gamma'}(1 + R^2)^{-1}_{\Gamma\Gamma''}$$
$$+ [R(1 + R^2)^{-1}]_{\Gamma\Gamma''}[R(1 + R^2)^{-1}]_{\Gamma\Gamma'}$$
$$+ i\{(1 + R^2)^{-1}_{\Gamma\Gamma'}[R(1 + R^2)^{-1}]_{\Gamma\Gamma''} - (1 + R^2)^{-1}_{\Gamma\Gamma''}[R(1 + R^2)^{-1}]_{\Gamma\Gamma'}\}.$$

Let $f_{\Gamma'\Gamma''}$ represent the product of the two matrix elements; if

$$f_{\Gamma'\Gamma''} = f_{\Gamma''\Gamma'},$$

which will later be shown to be true, then

$$\sum_{\Gamma'\Gamma''}(1 + R^2)^{-1}_{\Gamma\Gamma'}[R(1 + R^2)^{-1}]_{\Gamma\Gamma''} f_{\Gamma'\Gamma''} = \sum_{\Gamma''\Gamma'}(1 + R^2)^{-1}_{\Gamma\Gamma''}[R(1 + R^2)^{-1}]_{\Gamma\Gamma'} f_{\Gamma'\Gamma''},$$

having used the symmetry of f and interchanged the summation labels. With this result we see that the imaginary parts of the double sum over Γ' and Γ'' cancel each other! The real part of the R-dependent factors are the only factors which contain Γ, hence

$$\sum_{\Gamma}\{(1 + R^2)^{-1}_{\Gamma\Gamma'}(1 + R^2)^{-1}_{\Gamma\Gamma''} + [R(1 + R^2)^{-1}]_{\Gamma\Gamma''}[R(1 + R^2)^{-1}]_{\Gamma\Gamma'}\} = (1 + R^2)^{-1}_{\Gamma'\Gamma''}, \quad (2.146)$$

and the right-hand side of Eq. 2.145 can be written in the form

$$\sum_{M_L M_S m_1} \frac{2}{\pi} \sum_{\Gamma'\Gamma''} \langle\tilde{\Psi}(\Gamma')| \sum_i \nabla_i^{m_1} |\Phi(LS\Pi)\rangle \quad (1 + R^2)^{-1}_{\Gamma'\Gamma''} \times \langle\tilde{\Psi}(\Gamma'')| \sum_i \nabla_i^{m_1} |\Phi(LS\Pi)\rangle^* \quad (2.147)$$

We now proceed to evaluate these matrix elements in terms of the electronic orbitals, bound and free, and the geometric factors arising from the angular integrations and spin summations.

Problem 24. Show that if R is symmetric, its inverse is also symmetric Show that the real and imaginary parts of $(1 + iR)_{\alpha\beta}^{-1}$ are $(1 + R^2)_{\alpha\beta}^{-1}$ and $[R/(1 + R^2)]_{\alpha\beta}$, respectively. Prove the result quoted in Eq. 2.146.

When the first term in Eq. 2.142 is substituted into the matrix element, we have

$$\left\langle (N + 1)^{-1/2} \sum_{p=1}^{N+1} (-1)^{N+1-p} \sum_{\Gamma} \psi(\Gamma, \mathbf{X}\hat{x}_p) F_{\Gamma\Gamma'}(r_p) r_p^{-1} \sum_i \nabla_i^m \Phi(LS\Pi) \right\rangle.$$

We shall use the notation of Fano [53] to show that the basis functions ψ and Φ are constructed by vector coupling wave functions of properly antisymmetrized subshells, namely,

$$\psi(\Gamma, \mathbf{X}\hat{x}) = n(N_\lambda)^{-1/2} \sum_q (-1)^{P_q} \psi_u(q\Gamma\mathbf{X}\hat{x}_p)$$

$$= n(N_\lambda)^{-1/2} \sum_q (-1)^{P_q}\left[\left[\prod_\lambda (q_\lambda \mid nl_\lambda^{N_\lambda}\alpha_\lambda L_\lambda S_\lambda\}\right]^{\gamma} \times (p \mid k_\Gamma l_\Gamma \tfrac{1}{2}\}^\Gamma\right]. \quad (2.148)$$

where $(q_\lambda \mid nl_\lambda{}^{N\lambda}\alpha_\lambda L_\lambda S_\lambda\}$ represents the properly antisymmetrized wave function for the N_λ electrons in the subshell of nl_λ with subshell quantum numbers $\alpha_\lambda L_\lambda S_\lambda$ and electron labels denoted collectively by q_λ. The aggregate $q \equiv \{q_\lambda\}$ represents one distribution of the labels of the N electrons in the residual ion into groups of N_λ elements. The distribution of labels corresponding to normal order: $1, 2, 3, \ldots, N$, will be denoted by q_0. We take the vector product of the subshell wave functions to construct the residual ion function with quantum numbers denoted collectively by γ. The symbol $n(N_\lambda)$ is the number of possible distributions q of the N electrons in the residual atomic system after the photoelectron has been ejected. A similar expression to Eq. 2.148 can be written for Φ, which of course contains only bound orbitals for $(N + 1)$ electrons, not N electrons. The factor P_q is the number of permutations required to take the distribution q to normal order, q_0. Since the operator

$$\sum_{\alpha=1}^{N+1} \nabla_\alpha{}^m$$

is symmetric under the exchange of labels of any pair of particles, while Ψ and Φ are both antisymmetric under such interchange, then each of the $n(N_\lambda)$ terms in Σ_q will contribute equally, for each of the $N + 1$ values of p. Consequently, the contribution to the matrix element is

$$\sum_\Gamma \left[\frac{(N+1)n(N_\lambda)}{n(N_\lambda{}^\Phi)}\right]^{1/2} \sum_q (-1)^{P_q} \left\langle \psi_u(q_0\Gamma \mathbf{X}\,\hat{x}_{N+1}) \frac{F_{\Gamma\Gamma'}(r_{N+1})}{r_{N+1}} \times \sum_{\alpha=1}^{N+1} \nabla_\alpha{}^m \Phi_u(q, LS\Pi, \tau)\right\rangle \quad (2.149)$$

where $n(N_\lambda{}^\Phi)$ is the number of possible distributions q of the $N + 1$ electrons in the target atomic system, and the q in Eq. 2.149 now refers to the $N + 1$ labels distributed in Φ, not ψ!

We recall that the evaluation of Eq. 2.149 is to be carried out in the relaxed core approximation in which the bound orbitals of ψ for a given l are not orthogonal to the bound orbitals of Φ for that same l. In other words, the $N + 1$ radial integrals will be composed of the product of N overlap integrals of the form

$$\delta_{l_\psi l_\phi} \int_0^\infty dr P_{\rho_\psi}(r) P_{\rho_\phi}(r),$$

due to the spectator electrons and a single integral of the form

$$\int dr_\alpha P_{\rho_\psi}(r_\alpha) \left\{ r_\alpha \frac{d}{dr_\alpha} + \frac{l_{\rho_\phi}(l_{\rho_\phi} + 1) - l_{\rho_\psi}(l_{\rho_\psi} + 1)}{2r_\alpha} \right\} P_{\rho_\phi}(r)$$

depending on whether we evaluate the cross section in the dipole length, hence r_α, or dipole velocity forms. We have defined ρ_ψ to be the subshell containing the label α in $\psi_u(q_0)$; obviously, this subshell is uniquely defined by q_0 as is the corresponding distribution of labels among the N spectator electrons of ψ_u. The subshell containing the label α in $\phi_u(q)$ has been labelled ρ_ϕ; however, it is not specified uniquely like ρ_ψ, since we still have a sum over q!

Problem 25. Prove the commutator identity, $\mathbf{\nabla} = \frac{1}{2}[\nabla^2, \mathbf{r}]$. Using this identity, or otherwise, show

$$\int d\mathbf{r} R^*_{n'l'}(r) Y_{l'm'}(\hat{r})^* \nabla_\mu R_{nl}(r) Y_{lm}(\hat{r})$$

$$= \int d\hat{r} Y_{l'm'}(\hat{r})^* C_\mu{}^1 Y_{lm} \int r^2\, dr R^*_{n'l'} \left[\frac{1}{r}\frac{d}{dr} r + \frac{l(l+1) - l'(l'+1)}{2r}\right] R_{nl'}$$

where $C_\mu{}^1$ was defined in Eq. 2.63. See M. Rotenberg *et al.* [48].

In view of the lack of orthogonality between the orbitals of ψ and ϕ, the matrix element in Eq. 2.149 will have contributions from many more terms than the $\langle\psi|\, op\, |\psi\rangle$ matrix element encountered in the electron-scattering problem. We now develop the step-by-step reduction of (2.149). For a particular interacting electron label α, q_0 uniquely specifies the interacting subshell in ψ, ρ_ψ say, as well as the distribution of labels among the spectator electrons, namely $\bar{q}_0$. Within Σ_α, the interacting electron will be assigned to ρ_ψ, N_{ρ_ψ} times; each of these N_{ρ_ψ} terms will give the same contribution to the value for the matrix element, since we construct ψ such that it is antisymmetric under the interchange of any pair of labels within a subshell and Φ is antisymmetric under interchange of any pair of labels. Consequently, we need only to evaluate the matrix element, (2.149), once for each subshell in ψ; we shall use the label α_0 of the last electron in each subshell. In (2.149) we replace

$$\sum_\alpha \quad \text{with} \quad \sum_{\rho_\psi} N_{\rho_\psi}.$$

For each ρ_ψ, the label of the interacting electron is

$$\alpha_0 = \sum_{\lambda=1}^{\rho_\psi} N_\lambda.$$

We must now locate α_0 within each distribution q in the initial-state wave function. By inspection of (2.149) we know immediately that for a fixed ρ_ψ, the subshell containing α_0 in ϕ, ρ_ϕ say, cannot be arbitrarily any of the occupied subshells of ϕ! There is a selection rule due to ∇_μ having the rotational transformation properties of $Y_{1\mu}(\theta\phi)$. We must have $l_{\rho_\phi} = |l_{\rho_\psi} - 1|$ and $l_{\rho_\psi} + 1$.

In other words, for a fixed ρ_ψ, α_0 can only be found in those ρ_ϕ which are specified by this selection rule. Among all the distributions in Σ_q, there will be a set of them which assign α_0 to ρ_ϕ; the number in this set will be given by the number of distributions, $n(\bar{N}_\lambda)$, of the spectator labels $\bar{q} \equiv 1, 2, 3, \ldots, \alpha_0 - 1, \alpha_0 + 1, \ldots, N, N + 1$. In (2.149) we replace

$$\sum_q \quad \text{with} \quad \sum_{\rho_\phi(\rho_\psi)\cdot\bar{q}} \sum$$

where we have indicated the explicit dependence of ρ_ϕ on ρ_ψ.

Previously it has been emphasized that the radial orbitals of ψ and ϕ, for the same orbital quantum number l, are not assumed orthonormal to each other. However, we still have the orthogonality of the surface harmonics, $Y_{lm}(\theta\phi)$, to take into account when we evaluate the overlap integrals over the coordinates of the spectator electrons. That is to say, within $\bar{q}_0$ the fixed distribution of labels among the s-subshells imposes the constraint that these same labels can only be distributed among s-subshells in $\bar{q}$; similarly for the p, d, f, and so on subshells. For example, if $\bar{q}_0$ is $1s(1, 2)2s(3, 4)2p(5, 6, 7)$, then we must have only those $\bar{q}$ which assign labels 1, 2, 3, 4 to occupied s-subshells of ϕ. These conditions can be expressed by replacing

$$\sum_{\bar{q}} \quad \text{with} \quad \prod_{l=0}^{l^\phi} \left\{ \sum_{\bar{q}_l} \right\},$$

where l^ϕ is the occupied subshell with the highest orbital quantum number in ϕ. If $\bar{N}^l$ is the number of spectator electrons in $\bar{q}_0$ of ψ, in subshells with orbital angular momentum l, then we must have precisely this number of spectator l-electrons in $\bar{q}$, of ϕ; hence in actual fact we must replace

$$\sum_{\bar{q}} \quad \text{with} \quad \prod_{l=0}^{l^\phi} \left\{ \delta(\bar{N}^l, \bar{N}_\phi{}^l) \sum_{\bar{q}_l} \right\}$$

where $\bar{N}_{\lambda_\phi}{}^l$ is the number of spectator electrons in ϕ in subshells with orbital angular momentum l. If there are λ_l subshells in ϕ with orbital angular momentum l, then the number of distributions in $\Sigma\bar{q}_l$ will be

$$n(\bar{N}^l) = \frac{\bar{N}^l!}{\prod\limits_{\lambda_\phi=1}^{\lambda_l} (\bar{N}_{\lambda_\phi}{}^l!)}$$

where $\bar{N}_{\lambda_\phi}{}^l$ is the number of spectator electrons in the subshell λ_ϕ of ϕ with orbital quantum number l.

We can now rewrite (2.149) in the form

$$\sum_{\Gamma}\left[\frac{(N+1)n(N_\lambda)}{n(N_\lambda{}^\phi)}\right]^{1/2}\sum_{\rho_\psi}N_{\rho_\psi}\sum_{\rho_\phi(\rho\psi)}\prod_{l=0}^{l\phi}\left\{\delta(\bar{N}^l,\bar{N}_\phi{}^l)\sum_{\bar{q}_l}(-1)^{P_q}\right\}$$
$$\times\left\langle\psi_u(q_0\Gamma\mathbf{X}\hat{x}_{N+1})\,\frac{F_{\Gamma\Gamma'}(r_{N+1})}{r_{N+1}}\nabla_{\alpha_0}{}^m\phi_u(q,LS\Pi,\tau)\right\rangle. \quad (2.150)$$

To evaluate this matrix element we follow Fano [53] and separate out the interacting electron, label α_0, from ψ and ϕ, using coefficients of fractional parentage, namely,

$$\psi_u(q_0,\Gamma)=\sum_{\bar{S}_{\rho_\psi}\bar{L}_{\rho_\psi}}(l_{\rho_\psi}^{N_{\rho_\psi}}\alpha_{\rho_\psi}S_{\rho_\psi}L_{\rho_\psi}\,|\,\}l_{\rho_\psi}^{\bar{N}_{\rho_\psi}}\bar{\alpha}_{\rho_\psi}\bar{S}_{\rho_\psi}\bar{L}_{\rho_\psi},l_{\rho_\psi}(\alpha_0))\psi_{u_{\rho_\psi}}(\bar{q}_0\alpha_0,\Gamma),$$

and

$$\phi_u(q,LS\Pi)=\sum_{\bar{S}_{\rho_\phi}\bar{L}_{\rho_\phi}}(l_{\rho_\phi}^{N_{\rho_\phi}}\alpha_{\rho_\phi}S_{\rho_\phi}L_{\rho_\phi}\,|\,\}l_{\rho_\phi}^{\bar{N}_{\rho_\phi}}\bar{\alpha}_{\rho_\phi}\bar{S}_{\rho_\phi}\bar{L}_{\rho_\phi},l_{\rho_\phi}(\alpha_0))\phi_{u_{\rho_\phi}}(\bar{q}\alpha_0,\Gamma).$$

In general, $\bar{q}_0$ and $\bar{q}$ will not contain the same distribution of spectator electrons. Let $\beta\neq\alpha_0$ be the generic label for these spectators. We pick an allowed distribution in $\bar{q}$ from among the totality,

$$\prod_{l=0}^{l\phi}n(\bar{N}^l),$$

of such distributions which must be summed over in Eq. 2.150. We then let β range over its N possible labels, evaluating

$$\delta_{l_{\lambda_\psi}l_{\lambda_\phi}}\int_0^\infty dr_\beta P_{\lambda_\psi}(r_\beta)P_{\lambda_\phi}(r_\beta)$$

for each spectator label, where λ_ψ and λ_ϕ are the subshells in ψ and ϕ, respectively, where we located that particular label β. If $\lambda_\psi=\lambda_\phi$, then we proceed to the next β.

Those spectator electrons for which $\lambda_\psi\neq\lambda_\phi$ can be regarded as pseudo-interacting electrons and their generic label will be γ. In other words, a recoupling will be necessary to bring each γ to normal order in ϕ. Let γ be found in subshell $\sigma(\sigma')$ in the bra(ket). To begin this recoupling process, we must factor each γ-label from both the bra and the ket, using CFP's. The matrix element in (2.150) can be written now as

$$\sum_{\Gamma}[(N+1)\mathscr{n}(N_\lambda)/\mathscr{n}(N_\lambda{}^\phi)]^{1/2}\sum_{\rho_\psi}N_{\rho_\psi}\sum_{(\bar\alpha\bar S\bar L)_{\rho_\psi}}(l_{\rho_\psi}^{N_{\rho_\psi}}\alpha SL\,|\,\}l_{\rho_\psi}^{\bar N_{\rho_\psi}}\bar\alpha\bar S\bar L\,,l_{\rho_\psi}(\alpha_0))$$

$$\times\sum_{\rho_\phi}\prod_{l=0}^{l\phi}\delta(\bar N^l,\bar N_\phi{}^l)(l_{\rho_\psi}\|\nabla\|l_{\rho_\phi})(L1M_Lm\,|\,L_{\Gamma'}M_{\Gamma'})\sum_{(\bar\alpha\bar S\bar L)_{\rho_\phi}}$$

$$\times(l_{\rho_\phi}^{N_{\rho_\phi}}\alpha SL\,|\,\}l_{\rho_\phi}^{\bar N}\bar\alpha\bar S\bar L,l_{\rho_\phi}(\alpha_0))\prod_l\sum_{\bar q_l}(-1)^{P\bar q_l}\}\,\delta_{SS\Gamma}\,\delta_{M_SM_{S\Gamma}}$$

$$\times\prod_{\beta=1,\neq\alpha_0}^{N+1}\Bigg[\delta_{l_{\lambda\psi}l_{\lambda\phi}}\int_0^\infty dr_\beta P_{\lambda\psi}(r_\beta)P_{\lambda\phi}(r_\beta)$$

$$\times\sum_{L_\sigma\bar S_\sigma}\sum_{L_{\sigma'}\bar S_{\sigma'}}(l_\sigma{}^{N_\sigma}\alpha SL\,|\,\}l_\sigma{}^{\bar N_\sigma}\bar\alpha\bar S\bar L,l_\sigma(\gamma))(l_{\sigma'}^{N_{\sigma'}}\alpha SL\,|\,\}l_{\sigma'}^{\bar N_{\sigma'}}\bar\alpha\bar S\bar L,l_{\sigma'}(\gamma))\Bigg]$$

$$\times\langle\bar{\bar q}_0\{\gamma\}_\sigma\alpha_0;\ldots[\bar L_{\rho_\psi},(l_{\rho_\phi}1)l_{\rho_\psi}]L_{\rho_\psi}\ldots\bar L_{\rho_\phi}\ldots;L_\Gamma\,|\,\bar{\bar q}_0\{\gamma\}_{\sigma'}\alpha_0;\ldots[\bar L_{\rho_\phi}l_{\rho_\phi}]$$

$$L_{\rho_\phi}\ldots;L1;L_\Gamma)\rangle$$

$$\times\langle\ldots[\bar S_{\rho_\psi}\tfrac12(\alpha_0)]S_{\rho_\psi}\ldots\bar S_{\rho_\phi}\ldots;S_\Gamma\,|\ldots\bar S_{\rho_\psi}\ldots[\bar S_{\rho_\phi}\tfrac12(\alpha_0)]S_{\rho_\phi}\ldots;S\rangle$$

$$\equiv(L1M_Lm\,|\,L_{\Gamma'}\,M_{L_{\Gamma'}})\,\delta_{SS_{\Gamma'}}\,\delta_{M_SM_{S_{\Gamma'}}}G(\eta'L_{\Gamma'}S_{\Gamma'}\Pi_{\Gamma'};LS\Pi). \tag{2.151}$$

Equation 2.151 defines the real numbers G; η is the channel index (L_1S_1l), and $\bar{\bar q}_0$ denotes labels of the "genuine" spectator electrons in the matrix element. A computer program for evaluating the recoupling coefficients has been published by Burke [77]. The product over the spectator labels β involves recoupling for each label γ, an element of the set β, for which $\lambda_\psi\neq\lambda_\epsilon$, where $\sigma\equiv\lambda_\psi$ and $\sigma'\equiv\lambda_\epsilon$.

The contribution of the second term in Eq. 2.142 to the matrix element, that is,

$$\langle\Phi_\mu(\Gamma')|\sum_i\nabla_i{}^m|\Phi(LS\Pi)\rangle,$$

can be carried out in precisely the same way as that described above, with a bound orbital replacing $F(r_{N+1})$,

$$[(N+1)\mathscr{n}(N_\lambda)]^{1/2}\quad\text{replaced by}\quad\mathscr{n}(N_\lambda{}^\mu)^{1/2},$$

the number of ways of distributing the labels of $N+1$ electrons among the subshells of Φ_μ, and no analog to the sum over Γ appearing in Eq. 2.151. The result can be written as

$$\langle\Phi_\mu(\Gamma')|\sum_i\nabla_i{}^m|\Phi(LS\Pi)\rangle$$

$$=(L1M_Lm\,|\,L_{\Gamma'}M_{\Gamma'})G(\mu,L_{\Gamma'}S_{\Gamma'}\Pi_{\Gamma'};LS\Pi)\,\delta_{SS_{\Gamma'}}\,\delta_{M_SM_{S_{\Gamma'}}}. \tag{2.152}$$

When we combine (2.151) with (2.152) we have the matrix element is a real number given by

$$\langle\tilde{\Psi}(\Gamma')\sum_i \nabla_i^{m_1}\Phi(LS\Pi)\rangle = (L1M_Lm_1 \mid L_{\Gamma'}M_{\Gamma'})\,\delta_{SS_{\Gamma'}}\,\delta_{M_SM_{S_{\Gamma'}}}$$
$$\times\,[G(\eta' L_{\Gamma'}S_{\Gamma'}\Pi_{\Gamma'}; LS\Pi) + G(\mu L_{\Gamma'}S_{\Gamma'}\Pi_{\Gamma'}; LS\Pi)].$$

The product of the two matrix elements appearing in Eq. 2.145 is

$$f_{\Gamma'\Gamma''}(E, LS\Pi) \equiv (L1M_Lm_1 \mid L_{\Gamma'}M_{\Gamma'})(L1M_Lm_1 \mid L_{\Gamma''}M_{\Gamma''})\,\delta_{SS_{\Gamma'}}\,\delta_{SS_{\Gamma''}}$$
$$\times\,\delta_{M_SM_{S_{\Gamma'}}}\,\delta_{M_SM_{S_{\Gamma''}}}$$
$$\times\,[G(\eta' L_{\Gamma'}S_{\Gamma'}\Pi_{\Gamma'}; LS\Pi) + \sum_\mu C_{\Gamma'\mu}G(\mu L_{\Gamma'}S_{\Gamma'}\Pi_{\Gamma'}; LS\Pi)]$$
$$\times\,[G(\eta'' L_{\Gamma''}S_{\Gamma''}\Pi_{\Gamma''}; LS\Pi) + \sum_\mu C_{\Gamma''\mu}G(\mu L_{\Gamma''}S_{\Gamma''}\Pi_{\Gamma''}; LS\Pi)]$$
$$= f_{\Gamma''\Gamma'}(E, LS\Pi), \qquad (2.153)$$

as required earlier. Finally, we can rewrite Eq. 2.145 as

$$\sigma_v(E, LS\Pi) = \frac{4e^2\hbar^2}{m^2cv2(2L+1)(2S+1)}$$
$$\times \sum_{\Gamma'\Gamma''}(1 + R^2)^{-1}_{\Gamma'\Gamma''}f_{\Gamma'\Gamma''}(E, LS\Pi). \qquad (2.154)$$

2.3.4.2. *Differential Cross Sections.* The differential cross section for the absorption of a photon traveling along the z-axis with energy $h\nu$, resulting in the ejection of an electron with momentum $\hbar\mathbf{k}$, is given by

$$d\sigma(E, \hat{k}) = \frac{2\pi e^2\hbar^2}{m^2cv}\sum_{\gamma S_\Gamma M_{S_\Gamma}}[2(2S+1)(2L+1)]^{-1}\sum_{M_LM_Sm_1}$$
$$\times\,|\langle\overline{\Psi}(\gamma S_\Gamma M_{S_\Gamma}; \mathbf{k})|\sum_{i=1}^{N+1}\nabla_i^{m_1}|\Phi(LSM_LM_S\Pi)\rangle|^2\rho(E), \quad (2.155)$$

where $\gamma \equiv \{n_\Gamma L_1^{\ \Gamma}S_1^{\ \Gamma}M_{L_1\Gamma}\}$, and the density of final states is given by

$$\rho(E) = \frac{k\,d\hat{k}}{(2\pi)^3},$$

where $\overline{\Psi}$ is normalized per unit energy range. Neglecting the Coulomb complication at this time, we recall that we require the asymptotic form of the final state wave function for the photoelectron to be

$$e^{i\mathbf{k}\cdot\mathbf{r}} + \frac{f(\hat{k})e^{-ikr}}{r}.$$

Consequently, the $N + 1$ particle wave function must have the asymptotic form

$$\bar{\Psi}(\gamma S_\Gamma M_{S\Gamma}; \mathbf{k}) \underset{r\to\infty}{\sim} \psi(n_\Gamma L_1^\Gamma M_1^\Gamma (S_1^\Gamma \tfrac{1}{2}) S_\Gamma M_{S\Gamma}; \mathbf{X}\sigma_{N+1}) \left[e^{i\mathbf{k}\cdot\mathbf{r}} + \frac{f(\hat{k})e^{-ikr}}{r} \right] k^{-1/2}.$$

We note that this function is not in the Γ-representation that we have been working with previously. Therefore, we must make some transformations in order to isolate a radial function in a channel. To accomplish this goal we expand the photoelectric function in terms of Legendre polynomials

$$k^{-1/2}\left[e^{i\mathbf{k}\cdot\mathbf{r}} + f(\hat{k})\frac{e^{ikr}}{r} \right] = \sum_{l=0}^{\infty} a_l P_l(\hat{k}\cdot\hat{r})$$

$$= \sum_{l=0}^{\infty} a_l \frac{4\pi}{(2l+1)} \sum_{m=-l}^{l} Y_{lm}(\hat{k})^* Y_{lm}(\hat{r})$$

and then vector couple the functions

$$\psi Y_{lm} = \sum_{L_\Gamma M_\Gamma} (L_1^\Gamma l M_1^\Gamma m \mid L_\Gamma M_\Gamma) \bar{\Psi}(\Gamma; \mathbf{X}\hat{x}).$$

Combining these results together gives

$$\bar{\Psi}(\gamma S_\Gamma M_{S\Gamma}; \mathbf{k}) = \sum_{lm} Y_{lm}(\hat{k})^* \sum_{L_\Gamma M_\Gamma} (L_1^\Gamma l M_1^\Gamma m \mid L_\Gamma M_\Gamma) \Psi(\Gamma; \mathbf{X}\hat{x}) \quad (2.156)$$

where $\Psi(\Gamma; \mathbf{X}\hat{x})$ is given in terms of real functions by Eq. 2.142.

Equation 2.142 is substituted into Eq. 2.156 and the result is substituted into Eq. 2.155 to give

$$d\sigma(E, \hat{k})\left[\frac{2\pi e^2\hbar^2}{m^2 c\nu}\right]^{-1} = \sum_{\gamma S_\Gamma M_{S\Gamma}} [2(2S+1)(2L+1)]^{-1} \sum_{M_L M_S m_1}$$

$$\times \sum_{lml'm'} Y_{lm}(\hat{k}) Y_{l'm'}(\hat{k})^* \sum_{L_\Gamma M_\Gamma L_{\Gamma'} M_{\Gamma'}}$$

$$\times (L_1^\Gamma l M_1^\Gamma m \mid L_\Gamma M_\Gamma)(L_1^\Gamma l' M_1^\Gamma m' \mid L_{\Gamma'} M_{\Gamma'})$$

$$\times \Big\langle \left(\frac{2}{\pi}\right) \sum_{\Gamma_\alpha} (-i + R)^{-1}_{\Gamma\Gamma_\alpha} \bar{\Psi}(\Gamma_\alpha, E, \tau)$$

$$+ \sum_{\mu} C_{\Gamma\mu} \Phi_\mu(\Gamma, \tau) \mid \sum_i \nabla_i^{m_1} \mid \Phi(LS\Pi) \Big\rangle$$

$$\times \Big\langle \Phi(LS\Pi) \mid \sum_i \nabla_i^{m_1} \mid \left(\frac{2}{\pi}\right)^{1/2} \sum_{\Gamma_\beta} (-i + R)^{-1}_{\Gamma'\Gamma_\beta}$$

$$\bar{\Psi}(\Gamma_\beta, E, \tau) + \sum_{\nu} C_{\Gamma'\nu} \Phi_\nu(\Gamma', \tau) \Big\rangle$$

where

$$\Gamma \equiv \{nL_1{}^\Gamma S_1{}^\Gamma l L_\Gamma M_\Gamma S_\Gamma M_{S\Gamma}\Pi_\Gamma\} \quad \text{and} \quad \Gamma \equiv \{nL_1{}^\Gamma S_1{}^\Gamma l' L_\Gamma' M_\Gamma' S_\Gamma M_{S\Gamma}\Pi_{\Gamma'}\}. \tag{2.157}$$

When the two surface harmonics are combined into a single surface harmonic $Y_{PQ}(\hat{k})$, see Eq. 2.65, Eqs. 2.151, and so on are used for the matrix elements, then the sums over magnetic quantum numbers in the $\langle\tilde{\Psi} \mid \Phi\rangle^2$ terms are

$$\begin{aligned}
\sum(-1)^{m'}&(ll'm - m' \mid PQ)(L_1{}^\Gamma l M_1{}^\Gamma m \mid L_\Gamma M_\Gamma)(L_1{}^\Gamma l' M_1{}^\Gamma m' \mid L_\Gamma' M_\Gamma') \\
&\times (L1M_L m_1 \mid L_\alpha' M_\alpha')(-1)^{m_1}(L_{\beta'}1M_{\beta'} - m_1 \mid LM_L) \\
&= (-1)^{l+l'+L_\Gamma'-L_1{}^\Gamma}(2L_\Gamma + 1)[(2L_\Gamma' + 1)(2L_\Gamma + 1)(2L + 1)]^{1/2} \\
&\quad \times W(L_\Gamma' l' L_\Gamma l; L_1{}^\Gamma P)W(1L_\Gamma 1L_{\Gamma'}; LP)(-1)^{m_1} \\
&\quad \times (11m_1 - m_1 \mid PQ)\,\delta_{QO},
\end{aligned} \tag{2.158}$$

where we have used $L_\beta' = L_\Gamma$ and $L_\beta' = L_\Gamma'$. The fact that $Q \equiv 0$ means that the differential cross section for photoabsorption is independent of the azimuthal angle. From the Clebsch-Gordan coefficient we see that the only nonvanishing values of P are 0 and 2! We can see by inspection that the magnetic sums in Eq. 2.157 all give this result. Hence the differential cross section is

$$\begin{aligned}
d\sigma(E, \hat{k})\left[\frac{2\pi e^2\hbar^2}{m^2cv}\right]^{-1} &= \sum_\Gamma \sum_{l'L_\Gamma'} \sum_{Pm_1} (-1)^{l+l'+L_\Gamma'+L_1{}^\Gamma+m_1}(2L_\Gamma + 1) \\
&\quad \times [(2L_\Gamma' + 1)(2L_\Gamma + 1)(2L + 1)]^{1/2} \\
&\quad \times W(L_\Gamma' l' L_\Gamma l; L_1{}^\Gamma P)W(1L_\Gamma 1L_{\Gamma'}; LP) \\
&\quad \times \left[\frac{(2l + 1)(2l' + 1)}{4\pi(2P + 1)}\right]^{1/2} \\
&\quad \times (ll'00 \mid P0)(11m_1 - m_1 \mid P0)Y_{P0}(\hat{k}) \\
&\quad \times \Bigg\{\frac{2}{\pi}\sum_{\eta_\alpha\eta_\beta}(i + R)^{-1}_{\eta\eta_\alpha}(-i + R)^{-1}_{\eta'\eta_\beta} \\
&\quad \times G(\eta_\alpha L_\Gamma S_\Gamma \Pi_\Gamma; LS\Pi)G(\eta_\beta L_{\Gamma'} S_\Gamma \Pi_{\Gamma'}; LS\Pi) \\
&\quad + \left(\frac{8}{\pi}\right)^{1/2} \operatorname{Re}\sum_{\mu\eta_\alpha}(i + R)^{-1}_{\eta\eta_\alpha}G(\eta_\alpha L_\Gamma S_\Gamma \Pi_\Gamma; LS\Pi) \\
&\quad \times G(\mu L_{\Gamma'} S_\Gamma \Pi_{\Gamma'}; LS\Pi)C_{\Gamma'\mu} + \sum_{\mu\nu} C_{\Gamma\mu}C_{\Gamma'\nu} \\
&\quad \times G(\mu, L_\Gamma S_\Gamma \Pi_\Gamma; LS\Pi)G(\nu, L_{\Gamma'} S_\Gamma \Pi_{\Gamma'}; LS\Pi)\Bigg\} \\
&\equiv a_0 Y_{00}(\hat{k}) + a_2 Y_{20}(\hat{k}),
\end{aligned} \tag{2.159}$$

where $\eta' \equiv \{L_1 S_1 l'\}$.

Problem 26. Verify Eq. 2.158 by performing the sum over magnetic quantum numbers.

2.4. NUMERICAL METHODS FOR COUPLED DIFFERENTIAL EQUATIONS

Many scattering problems can be reduced to solving the system of M second ordinary differential equations

$$\frac{d^2F_\mu}{dr^2} = \sum_{\nu=1}^{M} U_{\mu\nu}(r)F_\nu(r), \qquad \mu = 1, 2, 3, \ldots, M, \tag{2.160}$$

with

$$U_{\mu\nu}(r) = U_{\nu\mu}(r) = \delta_{\mu\nu}\left\{-k_\nu^{\,2} + \frac{l_\nu(l_\nu+1)}{r^2} - \frac{2Z}{r}\right\} + v_{\mu\nu}(r) \tag{2.161}$$

see Eq. 2.13. We shall not specify whether or not $v_{\mu\nu}(r)$ is a short-range (exponential decay) potential. We recall that, in rydbergs, $k_\mu^{\,2} = \varepsilon - \varepsilon_\mu$, where ε is the total energy of the projectile plus target and ε_μ is the energy of the target in channel μ. When $k_\mu^{\,2}$ is positive, the radial functions F_μ will be oscillatory in the asymptotic domain, while if $k_1^{\,2}$ is such that

$$k_\nu^{\,2} = k_1^{\,2} - (|\varepsilon_1| - |\varepsilon_\nu|) < 0, \tag{2.162}$$

then the radial functions F_ν must decrease exponentially in the asymptotic domain in order to be physically significant. Consequently, for a given $k_1^{\,2}$, that is to say, for an incident beam of a given energy it is possible that this energy will be sufficient to excite some of the states allowed in the eigenfunction expansion, but insufficient to excite higher states. This will result in there being N_a positive (open channels) $k_\mu^{\,2}$ and $(M - N_a)$ negative (closed channels) $k_\nu^{\,2}$. We shall now develop a numerical method to solve such a system or ordinary differential equations.

2.4.1. Matching Algorithm [78]

We recall that for the open channels we must impose the asymptotic boundary condition, see Eq. 2.17,

$$\begin{gathered} F_\mu^{(\nu)}(r) \underset{r\to\infty}{\sim} k_\mu^{-1/2}\{\delta_{\mu\nu}\sin\theta_\mu + R_{\mu\nu}\cos\theta_\mu\}, \\ k_\mu^{\,2} > 0, \quad \mu \text{ and } \nu = 1, 2, \ldots, N_a, \end{gathered} \tag{2.163}$$

in the open channels. Since the task of the solution of the equations is to determine the elements of the R-matrix, we can hardly start with Eq. 2.163! What we do know is the form (2.163) is a linear superposition of the asymptotic forms of the regular and irregular Coulomb functions. Consequently, if we start with these functions, then our task is to construct the correct linear combination leading to (2.163). In other words, let us write the elements of the solution vector as

$$F_\mu(r) \underset{r\to\infty}{\sim} A_\mu \sin(\bar{\theta}_\mu) + B_\mu \cos(\bar{\theta}_\mu), \qquad \mu = 1, 2, \ldots, N_a, \qquad (2.164)$$

where $\bar{\theta}_\mu = k_\mu r + \eta_\mu \log 2k_\mu r$. If there are N_a open channels, then there will be $2N_a$ unknown constants A_μ and B_μ. In the closed channels, we see that the elements of the solution vector are

$$F_\tau(r) \underset{r\to\infty}{\sim} C_\tau e^{-\bar{\theta}_\tau}, \quad \text{where} \quad \bar{\theta}_\tau = -|k_\tau| r + \eta_\tau \log 2|k_\tau| r, \quad (2.165)$$

where there will be $(M - N_a)$ unknown coefficient constants C_τ.

Let us look at the question of unknown constants from another viewpoint. We have M second-order ordinary differential equations, each equation has two integration constants; consequently it is necessary to specify $2M$ constants at any value of the independent variable r before the numerical solution can begin. Let us look at the problem of starting the numerical integration of Eq. 2.160 at some $r = r_B$ way out in the asymptotic region. From Eq. 2.165 we know the $(M - N_a)$ integration constants

$$\left[\frac{dF_\tau/dr}{F_\tau}\right]_{r_B} = -|k_\tau|, \qquad (2.166)$$

which merely leaves $2M - (M - N_a) = (M + N_a)$ to be determined. The problem of starting the numerical integration at the origin, $r = 0$, is somewhat simpler, in that

$$F_\mu(r) \underset{r\to 0}{\sim} r^{l_\mu+1}, \qquad \mu = 1, 2, \ldots, M \qquad (2.167)$$

which means that all the elements of the solution vector vanish at the origin; this determines M of the $2M$ integration constants at $r = 0$, leaving M to be determined; for example the M slopes at $r = 0$.

At the origin, we can choose the M slopes in M linearly independent ways, for example by the simple technique

$$\left[\frac{dF_\mu(r)}{dr}\right]_{r=0}^{(\alpha)} = \delta_{\alpha\mu}, \qquad \alpha = 1, 2, \ldots, M. \qquad (2.168)$$

We are then in a position to numerically integrate the system of equations (2.160) step by step out from the origin, M *different times*. From these M linearly independent solutions, characterized by α, it will be possible to construct any solution. If this outward step-by-step integration was continued into the asymptotic domain, the numerical solutions would, in general, contain components of $e^{|k_\tau|r}$ as we got out to large r. These components would then swamp all knowledge of the oscillatory solutions and we should not have Eq. 2.166, the logarithmic derivative, satisfied. Consequently, the M outward integrations must be stopped at some r_0 small enough to ensure that k_μ^2 does not dominate in $U_{\mu\upsilon}(r)$. Let us denote the solutions and their derivatives so generated in this inner region $0 \leq r \leq r_0$ by $\mathcal{F}_\mu^{(\alpha)}(r)$ and $\mathcal{F}_\mu^{(\alpha)}(r)'$. As we have said, the linear superpositions

$$\sum_{\alpha=1}^{M} u_\alpha \begin{bmatrix} \mathcal{F}_\mu^{(\alpha)}(r) \\ \mathcal{F}_\mu^{(\alpha)}(r)' \end{bmatrix},$$

where the mixing coefficients are unknown as yet, can represent *any* solution in that region. Let us emphasize that such solutions will automatically satisfy the boundary conditions, (2.167), at the origin.

In order to generate a corresponding system of linearly independent solutions in the region $r_0 \leq r \leq r_B$ we consider Eqs. 2.164 and 2.165. These equations could be used to compute $F_\mu(r_B)$ and its derivative if we knew the $(M + N_a)$ coefficients A_μ, B_μ, and C_τ. This suggests that we generate $(M + N_a)$ different starting conditions at $r = r_B$, by setting A_μ, B_μ, C_τ equal to unity one at a time, the remaining coefficients zero. In order words, we set the elements of the $(M + N_a)$ dimensional vector

$$\begin{bmatrix} A_\mu \\ B_\mu \\ C_\tau \end{bmatrix}^\beta = \delta_{i\beta}, \qquad i = 1, 2, 3, \ldots, (M + N_a), \tag{2.169}$$

We can display this condition in the more explicit form

$$\begin{bmatrix} A_\mu \\ B_\mu \\ C_\tau \end{bmatrix}^1 = \begin{bmatrix} 1 \\ 0 \\ 0 \\ \cdot \\ \cdot \\ \cdot \\ 0 \end{bmatrix}, \quad \begin{bmatrix} A_\mu \\ B_\mu \\ C_\tau \end{bmatrix}^2 = \begin{bmatrix} 0 \\ 1 \\ 0 \\ 0 \\ \cdot \\ \cdot \\ \cdot \\ 0 \end{bmatrix}, \quad \begin{bmatrix} A_\mu \\ B_\mu \\ C_\tau \end{bmatrix}^3 = \begin{bmatrix} 0 \\ 0 \\ 1 \\ 0 \\ \cdot \\ \cdot \\ \cdot \\ 0 \end{bmatrix}, \ldots, \begin{bmatrix} A_\mu \\ B_\mu \\ C_\tau \end{bmatrix}^{M+N_a} = \begin{bmatrix} 0 \\ 0 \\ 0 \\ \cdot \\ \cdot \\ \cdot \\ 0 \\ 1 \end{bmatrix}.$$

With these $(M + N_a)$ linearly independent ways of assigning the coefficients, we can compute $\{F_\mu(r_B), F_\tau(r_B)\}$, and their derivatives, in $(M + N_a)$ linearly independent ways and so perform the step-by-step numerical integration of the system of differential equations from $r = r_B$ inward to $r = r_0$. The numerical solutions generated in this outer region will be denoted by $\mathcal{G}_\mu^{(\beta)}(r)$, $\beta = 1, 2, \ldots, (M + N_a)$, and any solution and its derivative can be given as a linear superposition

$$\sum_{\beta=1}^{M+N_a} \omega_\beta \begin{bmatrix} \mathcal{G}_\mu^{(\beta)}(r) \\ \mathcal{G}_\mu^{(\beta)}(r)' \end{bmatrix}, \quad \text{for} \quad r_0 \leq r \leq r_B \tag{2.170}$$

where the mixing coefficients ω_β are unknown as yet. It is clear that such solutions do indeed satisfy the boundary conditions (2.165) and do not have any component of the unwanted increasing exponential. However, if this inward solution was continued beyond r_0 and on to the origin, we should find that, in general, the solutions would not pass through the origin.

At the match point $r = r_0$, we shall impose the conditions that the inner and outer solutions, and their first derivatives, shall be continuous

$$\sum_{\alpha=1}^{M} u_\alpha \begin{bmatrix} \mathcal{F}_\mu \\ \mathcal{F}_\mu' \end{bmatrix}_{r_0}^{\alpha} = \sum_{\beta=1}^{M+N_a} \omega_\beta \begin{bmatrix} \mathcal{G}_\mu \\ \mathcal{G}_\mu' \end{bmatrix}_{r_0}^{\beta} \tag{2.171}$$

This will guarantee a smooth solution over the entire range, $0 \leq r \leq r_B$, indeed generates the solution which satisfies the boundary conditions both at the origin and asymptotically.

Equations 2.171 are a system of $2M$ equations in the $(2M + N_a)$ unknown parameters u_α and ω_β, in other words, at this time, we have an under-determined system. Our task now is to construct a further set of N_a equations. At r_B the form given in Eq. 2.163 must be constructed from Eq. 2.170. From Eq. 2.163 we see that there is not just a single solution vector $\{F_\mu(r)\}$ but N_a such solution vectors, denoted by ν in Eq. 2.163. Consequently, the set of mixing parameters used in Eq. 2.171 is not unique; there must be N_a such sets $\{u_\alpha, \omega_\beta\}^\nu$, $\nu = 1, 2, \ldots, N_a$. In other words, Eq. 2.171 and the additional N_a equations we have yet to find, must yield N_a sets of mixing parameters, so let us label Eq. 2.171 accordingly

$$\sum_{\alpha=1}^{M} u_\alpha^{(\nu)} \begin{bmatrix} \mathcal{F}_\mu \\ \mathcal{F}_\mu' \end{bmatrix}_{r_0}^{\alpha} = \sum_{\beta=1}^{M+N_a} \omega_\beta^{(\nu)} \begin{bmatrix} \mathcal{G}_\mu \\ \mathcal{G}_\mu' \end{bmatrix}_{r_0}^{\beta} \tag{2.172}$$

On the one hand, at $r = r_B$, the radial function part of Eq. 2.172, written out in detail for each μ is

$$\begin{aligned} &\omega_1^{(\nu)} \mathcal{G}_\mu^{\,1}(r_B) + \omega_2^{(\nu)} \mathcal{G}_\mu^{\,2}(r_B) + \cdots + \omega_{N_a+1}^{(\nu)} \mathcal{G}_\mu^{N_a+1}(r_B) + \cdots \\ &+ \omega_{2N_a+1}^{(\nu)} \mathcal{G}_\mu^{\,2N_a+1}(r_B) + \cdots + \omega_{M+N_a}^{(\nu)} \mathcal{G}_\mu^{M+N_a}(r_B) \rightarrow \omega_\mu^{(\nu)} \sin \bar{\theta}_\mu \\ &\qquad + \omega_{N_a+\mu}^{(\nu)} \cos \bar{\theta}_\mu, \end{aligned} \tag{2.173}$$

since by construction of the $(M + N_a)$ linearly independent solutions in the outer region, see Eq. 2.169, all the other $\mathcal{G}$'s are zero at r_B! On the other hand, from Eq. 2.163 we have

$$\begin{aligned} k_\mu^{-1/2}\{&\delta_{\mu\nu}[\sin \bar{\theta}_\mu \cos \phi_\mu - \cos \bar{\theta}_\mu \sin \phi_\mu] \\ &+ R_{\mu\nu}[\cos \bar{\theta}_\mu \cos \phi_\mu + \sin \bar{\theta}_\mu \sin \phi_\mu]\} \end{aligned}$$

where $\phi_\mu = \dfrac{l_\mu \pi}{2} - \sigma_{l\mu}$, which can be rewritten in the form

$$k_\mu^{-1/2}\{\sin \bar{\theta}_\mu[\delta_{\mu\nu} \cos \phi_\mu + R_{\mu\nu} \sin \phi_\mu] + \cos \bar{\theta}_\mu[-\delta_{\mu\nu} \sin \phi_\mu + R_{\mu\nu} \cos \phi_\mu]\}.$$

When the coefficients of $\sin \bar{\theta}_\mu$ and $\cos \bar{\theta}_\mu$ of this result are compared with Eq. 2.173 we obtain

$$\begin{aligned} \omega_\mu^{(\nu)} &= k_\mu^{-1/2}[\delta_{\mu\nu} \cos \phi_\mu + R_{\mu\nu} \sin \phi_\mu] \\ \omega_{\mu+N_a}^{(\nu)} &= k_\mu^{-1/2}[-\delta_{\mu\nu} \sin \phi_\mu + R_{\mu\nu} \cos \phi_\mu], \end{aligned} \tag{2.174}$$

from which we can readily derive the following pair of equations by subtraction and addition, respectively, and multiplying by either $\sin \phi_\mu$ or $\cos \phi_\mu$,

$$\omega_\mu^{(\nu)} \cos \phi_\mu - \omega_{\mu+N_a}^{(\nu)} \sin \phi_\mu = k_\mu^{-1/2}\delta_{\mu\nu}, \tag{2.175}$$

$$\omega_\mu^{(\nu)} \sin \phi_\mu + \omega_{\mu+N_a}^{(\nu)} \cos \phi_\mu = k_\mu^{-1/2}R_{\mu\nu}. \tag{2.176}$$

We see that Eq. 2.175 contains no additional unknowns to what we already have in the matching condition (2.172); furthermore, the system Eq. 2.175 is N_a in number, $\mu = 1, 2, 3, \ldots, N_a$ and gives us the set of extra equations we were looking for. In addition, we shall get N_a such families of extra equations depending on where we locate $k_\mu^{-1/2}$ on the right-hand side of Eq. 2.175. In summary, then, the $(M + N_a)$ equations in the $(M + N_a)$ unknowns ω_β and u_α can be displayed in the matrix form

$$\begin{array}{c} 1 \\ 2 \\ \cdot \\ \cdot \\ \cdot \\ M \\ 1 \\ 2 \\ \cdot \\ \cdot \\ \cdot \\ M \\ 1 \\ 2 \\ \cdot \\ \cdot \\ \cdot \\ N_a \end{array}
\left[\begin{array}{c|c} 1\ 2 \cdots M & 1\ 2\ 3 \cdots M+N_a \\ \mathscr{F}(r_0) & -\mathscr{G}(r_0) \\ \hline \mathscr{F}'(r_0) & -\mathscr{G}'(r_0) \\ \hline 0 & \cos\phi \quad \sin\phi \quad 0 \\ & 1\ 2 \cdots N_a, N_a+1, \cdots 2N_a \end{array}\right]
\left|\begin{array}{c} u_1 \\ u_2 \\ \cdot \\ \cdot \\ \cdot \\ u_M \\ \omega_1 \\ \omega_2 \\ \cdot \\ \cdot \\ \cdot \\ \cdot \\ \cdot \\ \omega_{M+N_a} \end{array}\right|
=
\left|\begin{array}{c} 0 \\ \hline 0 \\ 0 \\ \hline k^{-1/2} \end{array}\right|
\begin{array}{c} 1 \\ \cdot \\ \cdot \\ M \\ 1 \\ \cdot \\ M \\ 1 \\ \cdot \\ \cdot \\ \cdot \\ N_a \end{array}
\tag{2.177}$$

where the $\nu = 1$ solution is obtained by solving this system of equations with $k_1^{-1/2}$ in the $(2M + 1)$ position on the right-hand side, and where the $\nu = N_a$ solution, the final set of matching parameters, is obtained by having $k_{N_a}^{-1/2}$ in the $(2M + N_a)$th position on the right-hand side. So it is now clear how the N_a different sets of matching parameters $(u_1 u_2 \cdots u_M \omega_1 \cdots \omega_{M+N_a})^{(\nu)}$ are generated. With these known sets, it is a straightforward matter to calculate the elements of the R-matrix using Eq. 2.176, and hence the total cross sections. The fact that R must be symmetric provides a powerful check on the numerical procedures.

Problem 27. Given that the formula for the total cross sections for transitions from level $n'l_1'$ to level nl_1 of the hydrogen atom is (in units of πa_0^2)

$$Q(n'l_1' \to nl_1) = \sum_L \frac{(2L+1)}{k_{n'}^2} |T^L_{nl_1l_2,n'l_1'l_2'}|^2,$$

where $T = 2iR(I - iR)^{-1}$, write a computer program to calculate [79] the elastic $1s$-$1s$ and inelastic $1s$-$2s$ cross sections for electrons incident on atomic hydrogen, using the radial equations of Problem 7, at energies $k^2 = 1.0$, 1.21, 1.44, 2.25, and 4.0 for $L = 0$.

2.4.2. Constraints on the Radial Functions

In the independent particle model of the atom it is well known that the electrons are allocated to subshells which are identified by the quantum numbers first used for the states of the hydrogen atom, namely: $1s$, $2s$, $2p$, $3s$, $3p$, $3d$, ... The electrons are said to move in Hartree-Fock orbitals, which are radial functions and will be denoted by $P_{nl}(r)$ here, where

$$\int_0^\infty P_{nl}(r) P_{n'l'}(r)\, dr = \delta_{nn'}\, \delta_{ll'}. \tag{2.178}$$

This orthogonality condition implies, that in the atomic collision calculations we are going to concern ourselves with, the fact that not only are orbitals of different l orthogonal to one another, which is guaranteed by their associated angular functions alone, but even those orbitals with the same l, e.g., P_{1s} and P_{2s}, will be orthogonalized. In the analysis of electron scattering by many electron atomic systems, see Smith *et al.* [80], the algebra is considerably simplified if the radial functions F can be orthogonalized with respect to the orbitals P. We can look at the relationship between F and P in a somewhat different way: we have talked about electrons impinging on an atomic system containing N bound electrons. This is precisely the same as discussing an $(N + 1)$ electron atomic system in which one of the electrons, we do not know which one, since the electrons are indistinguishable from one another, is in a continuum orbital, which we have labelled F. Until now, all collision calculations have been performed assuming that the P_{nl} are known; a more accurate calculation would not make such an assumption but would generate its P_{nl} at the same time as its continuum orbitals. For F and P to be orthogonal we must impose the following integral condition, or constraint, on F

$$\int_0^\infty drP_{nl\lambda}(r)F_{\mu\nu}(r) = 0, \qquad l_\lambda = l_\mu \tag{2.179}$$

In other words, the set of mixing parameters, see Eq. 2.177, must be such as to generate this orthogonality as a by-product.

In Section 2.1.2 we presented one technique for deriving the radial equations. The more usual approach is to begin with a variational principle; this will be done in Section 2.5.1, and arrive at the expression, see Eq. 2.228,

$$\delta\left[\sum_{i,j}\int drF_{ik}D_{ij}F_{jl} - \frac{R_{kl}}{2}\right] = 0, \tag{2.180}$$

which is to be stationary with respect to variations in F such that

$$\delta F_{ij}(r) \underset{r\to\infty}{\sim} k_i^{-1/2}\,\delta R_{ij}\sin\theta_i. \tag{2.181}$$

The D-operators are defined later in Section 2.5.1.

The technique for accomplishing the orthogonalization of F with respect to bound orbitals rests on the use of Lagrange undetermined multipliers. Obviously, if $l_\mu \neq l_\lambda$, then the surface harmonics of $P(\mathbf{r})$ and $F(\mathbf{r})$ guarantee the orthogonality, so that the number of Lagrange multipliers, $\mathscr{L}$, required will have to be counted by taking each l_μ and comparing it with l_λ. Since we require $F_{\mu\nu}$ such that Eq. 2.179 is true, then we can add undetermined amounts of the overlap integrals to Eq. 2.180, that is, we have the trial value of the integral given by

$$I_{\alpha\beta} = \sum_{\gamma,\gamma'}\left\{\int_0^\infty drF_{\gamma\alpha}D_{\gamma\gamma'}F_{\gamma'\beta} + \sum_\lambda \mathscr{L}_{\lambda(\gamma')}\,\delta_{l\lambda l\gamma'}\int_0^\infty drP_{nl\lambda}F_{\gamma'\beta}\right.$$
$$\left. + \sum_\lambda \mathscr{L}_{\lambda(\gamma)}\,\delta_{l\lambda l\gamma}\int P_{nl\lambda}F_{\gamma\alpha}\,dr\right\} \tag{2.182}$$

The variation in $I_{\alpha\beta}$ due to a variation δF_{mn} in F_{mn} is

$$\int_0^\infty dr\,\delta F_{mn}(r)\left\{\delta_{n\alpha}\sum_{\gamma'}D_{m\gamma'}F_{\gamma'\beta} + \delta_{n\beta}\sum_{\gamma'}D_{m\gamma'}F_{\gamma'\alpha}\right\} + \tfrac{1}{2}\delta R_{\alpha\beta}$$
$$+ \sum_{\gamma\lambda}\delta_{l_\lambda l_\gamma}\mathscr{L}_{\lambda(\gamma)}\int_0^\infty drP_{nl\lambda}\,\delta F_{mn}\,\delta_{\gamma m}\,\delta_{\alpha n}$$
$$+ \sum_{\gamma'\lambda}\delta_{l_\lambda l_{\gamma'}}\mathscr{L}_{\lambda(\gamma')}\int_0^\infty drP_{nl\lambda}\,\delta F_{mn}\,\delta_{\gamma' m}\,\delta_{\beta n} = \tfrac{1}{2}\delta R_{\alpha\beta}.$$

Consequently we have

$$\int_0^\infty dr\, \delta F_{mn}(r)\Big\{\delta_{n\alpha}\Big[\sum_{\gamma'} D_{m\gamma'}F_{\gamma'\beta} + \sum_{\lambda} \delta_{l\lambda l_m}\mathscr{L}_{\lambda(n)}P_{nl\lambda}\Big]$$

$$+ \delta_{n\beta}\Big[\sum_{\gamma'} D_{m\gamma'}F_{\gamma'\alpha} + \sum_{\lambda} \delta_{l\lambda l_m}\mathscr{L}_{\lambda(n)}P_{nl\lambda}\Big]\Big\} = 0 \quad (2.183)$$

which can only be true for arbitrary variations δF_{mn}, provided the F's satisfy the Euler equations

$$\sum_{\gamma} D_{m\gamma}F_{\gamma\beta} + \sum_{\lambda} \delta_{l\lambda l_m}\mathscr{L}_{\lambda(m)}P_{nl\lambda}(r) = 0. \quad (2.184)$$

To solve this system of equations we must extend the numerical methods discussed previously. Let there be N_μ different Lagrange undetermined multipliers occurring in the system of equations (2.184). Therefore, in addition to the $(2M + N_a)$ mixing parameters already required in Eq. 2.172, we shall have a further N_μ unknowns. We can calculate a further N_μ independent solutions of the inhomogeneous system (2.184) by setting one of the $\mathscr{L}_{\lambda(m)}$ equal to unity and the remaining $(N_\mu - 1)$ undetermined multipliers zero. That is to say, we calculate N_μ such solutions, labelled $\mathscr{F}_\mu^{(\eta)}$, in the inner region and N_μ corresponding solutions, labelled $\mathscr{G}_\mu^{(\eta)}$, in the outer region, which are solutions of

$$\sum_{\gamma} D_{m\gamma}\mathscr{F}_\gamma^{(\eta)} + P_{nl_\eta}(r) = 0.$$

Let us multiply this system of equations by $\mathscr{L}_\eta$ and sum over η and we get

$$\sum_{\eta} \mathscr{L}_\eta \sum_{\gamma} D_{m\gamma}\mathscr{F}_\gamma^{(\eta)} + \sum_{\eta} \mathscr{L}_\eta P_{nl_\eta}(r) = 0,$$

or, when we take the number $\mathscr{L}_\eta$ through the differential operator $D_{m\gamma}$, we get

$$\sum_{\gamma} D_{m\gamma}\Big\{\sum_{\eta} \mathscr{L}_\eta \mathscr{F}_\gamma^{(\eta)}\Big\} + \sum_{\eta} \mathscr{L}_\eta P_{nl_\eta} = 0.$$

Furthermore, if we add the homogeneous solutions (2.171) in the form

$$\sum_{\gamma} D_{m\gamma}\left\{\begin{matrix} \sum_{\alpha} u_\alpha^{(\nu)}\mathscr{F}_\gamma^{\alpha} \\ \sum_{\beta} \omega_\beta^{(\nu)}\mathscr{G}_\gamma^{\beta} \end{matrix}\right\} = 0$$

to this result we obtain

$$\sum_\gamma D_{m\gamma}\begin{Bmatrix} \sum_\alpha u_\alpha^{(\nu)}\mathscr{F}_\gamma^{\ \alpha} + \sum_{\eta=1}^{N\mu}\mathscr{L}_\eta\mathscr{F}_\gamma^{\ \eta} \\ \sum_\beta \omega_\beta^{(\nu)}\mathscr{G}_\gamma^{\ \beta} + \sum_{\eta=1}^{N\mu}\mathscr{L}_\eta\mathscr{G}_\gamma^{\ \eta} \end{Bmatrix} + \sum_\eta \mathscr{L}_\eta P_{nl_\eta}(r) = 0, \qquad (2.185)$$

which is Eq. 2.184, the equation we set out to solve, provided

$$\sum_{\alpha=1}^{M} u_\alpha^{(\nu)}\mathscr{F}_\gamma^{\ \alpha}(r) + \sum_{\eta=1}^{N_\mu}\mathscr{L}_\eta\mathscr{F}_\gamma^{\ \eta}(r), \qquad 0 \le r \le r_0$$

and

$$\sum_{\alpha=1}^{M+N_a} \omega_\alpha^{(\nu)}\mathscr{G}_\gamma^{\ \alpha}(r) + \sum_{\eta=1}^{N_\mu}\mathscr{L}_\eta\mathscr{G}_\gamma^{\ \eta}(r), \qquad r_0 \le r \le r_B$$

are the solutions in the inner and outer regions, respectively, where η should carry a superscript ν and is the same in both outer and inner regions. The matching condition at r_0 is simply

$$\sum_{\alpha=1}^{M} u_\alpha^{\ \nu}\begin{bmatrix}\mathscr{F}_\gamma^{\ \alpha} \\ \mathscr{F}_\gamma^{\ \alpha'}\end{bmatrix}_{r_0} + \sum_{\eta=1}^{N_\mu}\mathscr{L}_\eta\begin{bmatrix}\mathscr{F}_\gamma^{\ \eta} \\ \mathscr{F}_\gamma^{\ \eta'}\end{bmatrix}_{r_0} = \sum_{\alpha=1}^{M+N_a}\omega_\alpha^{\ \nu}\begin{bmatrix}\mathscr{G}_\gamma^{\ \alpha} \\ \mathscr{G}_\gamma^{\ \alpha'}\end{bmatrix}_{r_0} + \sum_{\eta=1}^{N_\mu}\mathscr{L}_\eta\begin{bmatrix}\mathscr{G}_\gamma^{\ \eta} \\ \mathscr{G}_\gamma^{\ \eta'}\end{bmatrix}_{r_0}$$

$$(2.186)$$

which is $2M$ equations, $\gamma = 1, 2, \ldots, M$, in $(2M + N_a + N_\mu)$ unknowns. As before, additional N_a equations are given by Eq. 2.175. The remaining N_μ equations are given by the constraints Eq. 2.179 themselves,

$$\delta_{l_\lambda l_\gamma}\left[\int_0^{r_0} drP_{nl\lambda}(r)\left\{\sum_{\alpha=1}^{M} u_\alpha^{\ \nu}\mathscr{F}_\gamma^{\ \alpha} + \sum_{\eta=1}^{N_\mu}\mathscr{L}_\eta\mathscr{F}_\gamma^{\ \eta}\right\}\right.$$
$$\left. + \int_{r_0}^{\infty} drP_{nl\lambda}(r)\left\{\sum_{\beta=1}^{M+N_a}\omega_\beta^{\ \nu}\mathscr{G}_\gamma^{\ \beta} + \sum_{\eta=1}^{N_\mu}\mathscr{L}_\eta\mathscr{G}_\gamma^{\ \eta}\right\}\right] = 0, \qquad \gamma = 1, 2, \ldots, N_a.$$

The Kronecker delta sorts out the N_μ orthogonality relations. This result can be rewritten in a form more obvious for correlating with Eq. 2.186, namely

$$\sum_{\alpha=1}^{M} u_\alpha^{(\nu)}\int_0^{r_0} drP_{nl\lambda}\mathscr{F}_\gamma^{\ \alpha} + \sum_{\beta=1}^{M+N_a}\omega_\beta^{(\nu)}\int_{r_0}^{\infty} drP_{nl\lambda}\mathscr{G}_\gamma^{\ \beta}$$
$$+ \sum_{\eta=1}^{N_\mu}\mathscr{L}_\eta^{(\nu)}\left\{\int_0^{r_0} drP_{nl\lambda}\mathscr{F}_\gamma^{\ \eta} + \int_{r_0}^{\infty} drP_{nl\lambda}\mathscr{G}_\gamma^{\ \eta}\right\} = 0, \qquad (2.187)$$

where all the integrals can be evaluated once the linearly independent solutions of the equations have been generated. The matrix form of the matching equations, analogous to Eq. 2.177, are now the extended set

$$
\begin{array}{c|c:c:c||c||c||c|}
 & 1\,2\cdots M & M+1\cdots M+N_\mu & 1\,2\,3\cdots M+N_a & & & \\
\begin{matrix}1\\2\\ \cdot\\ \cdot\\ M\end{matrix} & \mathscr{F}^{\alpha} & \mathscr{F}^{\eta}-\mathscr{G}^{\eta} & \mathscr{G}^{\beta} & \begin{matrix}u_1\\u_2\\ \cdot\\ u_M\end{matrix} & & 0 \\
\hdashline
\begin{matrix}1\\2\\ \cdot\\ \cdot\\ M\end{matrix} & \mathscr{F}^{\alpha'} & \mathscr{F}^{\eta'}-\mathscr{G}^{\eta'} & \mathscr{G}^{\beta'} & \begin{matrix}\mathscr{L}_1\\ \mathscr{L}_2\\ \cdot\\ \cdot\end{matrix} & & 0 \\
\hdashline
\begin{matrix}1\\2\\ \cdot\\ \cdot\\ N_\mu\end{matrix} & \int_0^{r_0} drP\mathscr{F}_{\gamma}{}^{\alpha} & \int_0^{r_0} drP\mathscr{F}_{\gamma}{}^{\eta}+\int_{r_0}^{\infty} drP\mathscr{G}_{\gamma}{}^{\eta} & \int_{r_0}^{\infty} drP\mathscr{G}_{\gamma}{}^{\eta} & \begin{matrix}\cdot\\ \cdot\\ \mathscr{L}_{N_\mu}\\ \omega_1\\ \omega_2\end{matrix} & = & 0 \\
\hdashline
\begin{matrix}1\\2\\ \cdot\\ \cdot\\ N_a\end{matrix} & 0 & 0 & \begin{matrix}\cos\phi & \sin\phi & 0\\ 12\cdots N_a, & N_a+1\cdots 2N_a & \end{matrix} & \begin{matrix}\cdot\\ \cdot\\ \cdot\\ \omega_{M+N_a}\end{matrix} & & k_\nu^{-1/2}
\end{array}
\qquad (2.188)
$$

where the blocks containing integrals are meant to represent collectively elements such as

$$
B_{2M+\gamma',\alpha} \equiv \int_0^{r_0} drP_{nl\lambda}\mathscr{F}_{\gamma}^{\ \alpha}, \qquad \gamma = 1, 2, \ldots, N_\mu; \quad \alpha = 1, 2, \ldots, M
$$

where **B** is called the matching matrix.

The physical interpretation of imposing orthogonality is that the continuum electron has no radial component in the discrete subshells. That is to say, if electrons are being scattered from atomic systems with open subshells there is no possibility of the electron being captured into these subshells.

This is an obvious deficiency on the part of the trial function assumed in Eq. 2.140. In order to make allowance for the $(N + 1)$ electron configurations in which all orbitals are discrete, extra terms must be added. However, these extra terms must have the same $LS\Pi$ as Ψ_t. If we let μ denote the incomplete subshells included in the eigenfunction expansion, and $\Phi_\mu(LS\Pi; \mathbf{x}_1 \cdots \mathbf{x}_{N+1})$ denote the stationary eigenstates of the $(N + 1)$ electron system obtained by assigning an electron to each of the open subshells μ, then the trial function becomes, see Eq. 2.142

$$\Psi_t^{\alpha} = \sum_{\beta} \psi_\beta(\mathbf{X}, \hat{x}_{N+1}) \frac{F_\beta^{\alpha}(r_{N+1})}{r_{N+1}} + \sum_{\mu} C_\mu^{\alpha} \Phi_\mu(LS\Pi; \mathbf{X}\mathbf{x}_{N+1}), \quad (2.189)$$

where the coefficients C_μ^{α} are completely arbitrary. It is clear by inspection, that when the trial function (2.189) is used to calculate the numbers $I_{\alpha\beta}$, terms linear and quadratic in C appear in the analog to (2.180), namely the terms, see Eq. 2.228,

$$\sum_{\gamma',\mu} C_\mu^{\alpha} \int dr V_{\mu,\gamma'} F_{\gamma'}^{\beta} + \sum_{\gamma'\mu} C_\mu^{\beta} \int dr V_{\mu'\gamma} F_\gamma^{\alpha} + \sum_{\mu,\nu} C_\mu^{\alpha} C_\nu^{\beta} A_{\mu\nu}.$$

have to be added to Eq. 2.182. For variations of F_{mn}, subject to Eq. 2.181, we obtain the system of equations

$$\sum_{\gamma} D_{m\gamma} F_\gamma^{(\beta)} + \sum_{\eta} \mathscr{L}_\eta P_{nl\eta} + \sum_{\mu} C_\mu^{\beta} V_{\mu,m} = 0. \quad (2.190)$$

Variations of $I_{\alpha\beta}$ with respect to C_ν^{m} leads to the following conditions for determining the C's if the F's are known:

$$\sum_{\nu} A_{\mu\nu} C_\nu^{\beta} + \sum_{\gamma} \int_0^{\infty} V_{\mu,\gamma} F_\gamma^{\beta}\, dr = 0 \quad (2.191)$$

where $A_{\mu\nu}$ are essentially the matrix elements of the Hamiltonian calculated using Φ, and are known; $V_{\mu,\gamma}$ are known functions of r, since they are integrals over the discrete orbitals.

Let there be N_V unknown coefficients C_μ appearing in the system of second-order ordinary differential equations (2.190). The technique for evaluating these C_μ is a straightforward extension of the algorithm described above for the Lagrange undetermined multipliers $\mathscr{L}_\mu$. We now have $(2M + N_a + N_\mu + N_V)$ unknown parameters in the problem. In addition to the $(M + N_\mu)$ solutions generated in the inner regions with $C_\mu \equiv 0$, we generate N_V particular solutions to the inhomogeneous system (2.190) by setting each of the C_μ equal to unity and the remainder zero. Let N_I denote the total number of inner solutions, $N_I = M + N_\mu + N_V$. In addition to the $(M + N_a + N_\mu)$ solutions generated in the outer region with $C_\mu \equiv 0$, we generate a further N_V particular solution to the inhomogeneous system, setting $C_\mu{}^\varepsilon = \delta_{\mu\varepsilon}$, $\varepsilon = 1, 2, \ldots, N_V$. We can prove that the linear combinations of the linearly independent solutions will be solutions of Eq. 2.190 with the matching condition given by

$$\sum_{\alpha=1}^{M} u_\alpha^{(v)} \begin{pmatrix} \mathscr{F}_\gamma{}^\alpha \\ \mathscr{F}_\gamma{}^{\alpha\prime} \end{pmatrix}_{r_0} + \sum_{\eta=1}^{N\mu} \mathscr{L}_\eta^{(v)} \begin{pmatrix} \mathscr{F}_\gamma{}^\eta \\ \mathscr{F}_\gamma{}^{\eta\prime} \end{pmatrix}_{r_0} + \sum_{\varepsilon=1}^{N_V} C_\varepsilon^{(v)} \begin{pmatrix} \mathscr{F}_\gamma{}^\varepsilon \\ \mathscr{F}_\gamma{}^{\varepsilon\prime} \end{pmatrix}_{r_0}$$

$$= \sum_{\beta=1}^{M+N_a} \omega_\beta^{(v)} \begin{pmatrix} \mathscr{G}_\gamma{}^\beta \\ \mathscr{G}_\gamma{}^{\beta\prime} \end{pmatrix}_{r_0} + \sum_{\eta=1}^{N\mu} \mathscr{L}_\eta^{(v)} \begin{pmatrix} \mathscr{G}_\gamma{}^\eta \\ \mathscr{G}_\gamma{}^{\eta\prime} \end{pmatrix}_{r_0} + \sum_{\varepsilon=1}^{N_V} C_\varepsilon^{(v)} \begin{pmatrix} \mathscr{G}_\gamma{}^\varepsilon \\ \mathscr{G}_\gamma{}^{\varepsilon\prime} \end{pmatrix}_{r_0} \qquad (2.192)$$

which is $2M$ equations in $(2M + N_a + N_\mu + N_V)$ unknowns. As before, a further N_a equations are supplied by Eq. 2.175, which insists that in the asymptotic domain the coefficient of the sine function is $\delta_{\mu\gamma}$, and another N_μ equations are given by extending the set (2.187) to include the ε solutions, that is,

$$\sum_{\alpha=1}^{M} u_\alpha^{(v)} \int_0^{r_0} dr P_\eta \mathscr{F}_\eta{}^\alpha + \sum_{\beta=1}^{M+N_a} \omega_\beta^{(v)} \int_{r_0}^{\infty} dr P_\eta \mathscr{G}_\eta{}^\beta$$
$$+ \sum_{\eta'=1}^{N\mu} \mathscr{L}_{\eta'}^{(v)} \left\{ \int_0^{r_0} dr P_\eta \mathscr{F}_\eta{}^{\eta'} + \int_{r_0}^{\infty} dr P_\eta \mathscr{G}_\eta{}^{\eta'} \right\}$$
$$+ \sum_{\varepsilon=1}^{N_V} C_\varepsilon^{(v)} \left\{ \int_0^{r_0} dr P_\eta \mathscr{G}_\eta{}^\varepsilon + \int_{r_0}^{\infty} dr P_\eta \mathscr{G}_\eta{}^\varepsilon \right\} = 0, \qquad (2.193)$$

where $\eta = 1, 2, \ldots, N_\mu$.

The remaining N_V equations are derived by substituting the linear combination of independent solutions into Eq. 2.191, giving for $\mu = 1, 2, 3, \ldots, N_V$:

$$\sum_\beta A_{\mu\beta} C_\beta^{(v)} + \sum_{\alpha=1}^{M} u_\alpha^{(v)} \sum_\gamma \int_0^{r_0} dr V_{\mu,\gamma} \mathscr{F}_\gamma{}^\alpha + \sum_{\beta=1}^{M+N_a} \omega_\beta^{(v)} \sum_\gamma \int_{r_0}^{\infty} dr V_{\mu,\gamma} \mathscr{G}_\gamma^\alpha$$
$$+ \sum_{\eta'=1}^{N\mu} \mathscr{L}_{\eta'}^{(v)} \left\{ \sum_\gamma \int_0^{r_0} V_{\mu,\gamma} \mathscr{F}_\gamma{}^{\eta'}\, dr + \sum_\gamma \int_{r_0}^{\infty} V_{\mu,\gamma} \mathscr{G}_\gamma{}^{\eta'}\, dr \right\}$$
$$+ \sum_{\varepsilon=1}^{N_V} C_\varepsilon^{(v)} \left\{ \sum_\gamma \int_0^{r_0} dr V_{\mu,\gamma} \mathscr{F}_\gamma{}^\varepsilon + \sum_\gamma \int_{r_0}^{\infty} dr V_{\mu,\gamma} \mathscr{G}_\gamma{}^\varepsilon \right\} = 0 \qquad (2.194)$$

The matrix form of the matching equations, which extends the set, Eq. 2.188, are

$$
\begin{array}{c|c:c:c:c||c||c|}
 & 1\,2\cdots M & M+1\cdots M+N_\mu & 1\,2\cdots N_V & 1\,2\,3\cdots M+N_a & & \\
\begin{matrix}1\\ \vdots\\ M\end{matrix} & \mathcal{F}^{\alpha} & \mathcal{F}^{\eta}-\mathcal{G}^{\eta} & \mathcal{F}^{\varepsilon}-\mathcal{G}^{\varepsilon} & \mathcal{G}^{\beta} & \begin{matrix}u_1\\ u_2\\ \vdots\\ u_M\end{matrix} & 0 \\
\hdashline
\begin{matrix}1\\ \vdots\\ M\end{matrix} & \mathcal{F}^{\alpha'} & \mathcal{F}^{\eta'}-\mathcal{G}^{\eta'} & \mathcal{F}^{\varepsilon'}-\mathcal{G}^{\varepsilon'} & \mathcal{G}^{\beta'} & \begin{matrix}\mathcal{L}_1\\ \mathcal{L}_2\\ \vdots\end{matrix} & 0 \\
\hdashline
\begin{matrix}1\\ \vdots\\ N_\mu\end{matrix} & \int_0^{r_0} P_\eta \mathcal{F}_\eta^{\alpha} dr & \begin{matrix}\int_0^{r_0} P_\eta \mathcal{F}_\eta^{\eta'} dr\\ +\\ \int_{r_0}^{\infty} P_\eta \mathcal{G}_\eta^{\eta'} dr\end{matrix} & \begin{matrix}\int_0^{r_0} P_\eta \mathcal{F}_\eta^{\varepsilon} dr\\ +\\ \int_{r_0}^{\infty} P_\eta \mathcal{G}_\eta^{\varepsilon} dr\end{matrix} & \int_r^{\infty} P_\eta \mathcal{G}_\eta^{\beta} dr & \begin{matrix}\mathcal{L}_{N_\mu}\\ C_1\\ C_2\\ \vdots\end{matrix} & 0 \\
\hdashline
\begin{matrix}1\\ \vdots\\ N_V\end{matrix} & \int_0^{r_0} V\mathcal{F}^{\alpha} dr & \begin{matrix}\int_0^{r_0} V\mathcal{F}^{\eta} dr\\ +\\ \int_{r_0}^{\infty} V\mathcal{G}^{\eta} dr\end{matrix} & \begin{matrix}A\\ +\\ \int_0^{r_0} V\mathcal{F}^{\varepsilon} dr\\ +\\ \int_{r_0}^{\infty} V\mathcal{G}^{\varepsilon} dr\end{matrix} & \int_r^{\infty} V\mathcal{G}^{\beta} dr & \begin{matrix}\vdots\\ C_{N_V}\\ \omega_1\\ \omega_2\end{matrix} & 0 \\
\hdashline
\begin{matrix}1\\ \vdots\\ N_a\end{matrix} & 0 & 0 & \begin{matrix}\ddots\\ \cos\phi\ \sin\phi\quad 0\\ \ddots\end{matrix} & & \begin{matrix}\vdots\\ \omega_{M+N_a}\end{matrix} & k_\nu^{-1/2}
\end{array}
= \quad (2.195)
$$

As before, $\nu = 1, 2, \ldots, N_a$ different solutions to the vector of matching coefficients are generated by varying the location of $k_\nu^{-1/2}$; knowing $\omega_\mu^{(\nu)}$ we can calculate $R_{\mu\nu}$!

2.4.3. Separable Kernels in Integro-Differential Equations

Those scattering problems which involve rearrangement of the constituent particles lead to integro-differential equations, rather than equations of the form (2.160). That is to say, the radial equations describing the scattering can be written as

$$\frac{d^2F_\mu}{dr^2} = \sum_{\nu=1}^{M}\left[U_{\mu\nu}(r)F_\nu(r) + \int k_{\mu\nu}(r, r')F_\nu(r')\,dr'\right] \tag{2.196}$$

where k is called the kernel of the system, and is generally given by

$$k_{\mu\nu}(r, r') = \iint R_\mu(\mathbf{r}')^* V(\mathbf{r}, \mathbf{r}')R_\nu(\mathbf{r})Y_{lm}(\theta'\phi')\,d\Omega\,d\Omega', \tag{2.197}$$

where $R_i(\mathbf{r})$ represents the bound orbitals of the particles of the target system calculated according to an independent particle model, $V(r, r')$ represents the interaction between the incoming projectile and the constituent particles of the target system, and Y_{lm} is the angular part of the radial function F_ν. For example, for electron and positron scattering from atomic systems,

$$V(\mathbf{r}, \mathbf{r}')_{\text{elec}} = |\mathbf{r} - \mathbf{r}'|^{-1},$$

while for nucleon scattering by nuclei, the internucleon potential is often given by its Yukawa form

$$V(\mathbf{r}, \mathbf{r}')_{\text{nuc}} = \frac{e^{-\alpha|\mathbf{r}-\mathbf{r}'|}}{|\mathbf{r} - \mathbf{r}'|}$$

These two examples of particle-particle interactions belong to a class of potentials, including [81] $\exp[-\alpha|\mathbf{r} - \mathbf{r}'|]$ and $|\mathbf{r} - \mathbf{r}'|\exp[-\alpha|\mathbf{r} - \mathbf{r}'|]$, which lead to kernels of the type which we have called separable.

Problem 28. In the static-exchange approximation to the scattering of electrons by hydrogen-like systems, that is, including electron exchange, the overall spatial wave function for the system is given by

$$\psi(\mathbf{r}_1, \mathbf{r}_2) = \phi(\mathbf{r}_1)F(\mathbf{r}_2) \pm \phi(\mathbf{r}_2)F(\mathbf{r}_1),$$

where ϕ is the $1s$ wave function. Derive the explicit form of the kernel for this system.

It is well-known [81] that the above potentials can be expanded in terms of Legendre polynomials, namely, for $r_{12} = |\mathbf{r}_1 - \mathbf{r}_2|$,

$$\frac{e^{-\alpha r_{12}}}{r_{12}} = \sum_{\lambda=0}^{\infty}\frac{(2\lambda + 1)}{(r_1r_2)^{1/2}} I_{\lambda+1/2}(\alpha r_<)K_{\lambda+1/2}(\alpha r_>)P_\lambda(\hat{r}_1 \cdot \hat{r}_2), \tag{2.198}$$

where $I_{\gamma+1/2}$ and $K_{\gamma+1/2}$ are the modified Bessel functions, whose arguments include the symbols $r_>$ and $r_<$, which represent the greater of r_1 and r_2, and the lesser of r_1 and r_2, respectively. When Eqs. 2.198 and 2.197 are substituted back into Eq. 2.196 we have the "exchange" term given by

$$\sum_{\lambda=0}^{\infty} \frac{(2\lambda+1)}{(r_1 r_2)^{1/2}} \iint d\Omega_1\, d\Omega_2 Y_{l_\mu m_\mu}(\theta_2\phi_2)^* P_\lambda(\hat{r}_1 \cdot \hat{r}_2) Y_{l_\nu m_\nu}(\theta_1\phi_1) Y_{lm}(\theta_2\phi_2)$$
$$\times \int_0^\infty p_\mu(r_2) I_{\lambda+1/2}(\alpha r_<) K_{\lambda+1/2}(\alpha r_>) p_\nu(r_1) F_\nu(r_2)\, dr_2,$$

where p represents the bound radial orbitals of particles in the target. The angular integrations can always be performed to yield a known geometrical coefficient, g_λ say. By analogy with the well-known Slater integrals, we shall define "Yukawa integrals" by

$$\chi_\lambda^{(\alpha)}(p_\mu, F_\nu; r_1) \equiv r_1 \int_0^{r_1} dr_2 p_\mu(r_2) F_\nu(r_2) \frac{I_{\lambda+1/2}(\alpha r_2) K_{\lambda+1/2}(\alpha r_1)}{(r_1 r_2)^{1/2}}$$
$$+ r_1 \int_{r_1}^\infty dr_2 p_\mu(r_2) F_\nu(r_2) \frac{I_{\lambda+1/2}(\alpha r_1) K_{\lambda+1/2}(\alpha r_2)}{(r_1 r_2)^{1/2}} \tag{2.199}$$

Consequently, the radial equations can be written in the form

$$\frac{d^2 F_\eta}{dr^2} = \sum_{\nu=1}^{M} \left[U_{\mu\nu}(r) F_\nu(r) + \sum_{\lambda=0}^{\infty} g_\lambda(\mu\nu) \chi_\lambda^{(\alpha)}(p_\mu F_\nu; r) \frac{p_\nu(r)}{r} \right], \tag{2.200}$$

where $\chi(r)$ is an unknown function of r. In other words, since χ needs F before it can be evaluated, and the differential equations for F need χ before they can be solved, then we seem to be locked into an iterative procedure.

The purpose of this section is to show how $\chi(r)$ satisfies a second-order ordinary differential equation involving F; consequently, if there are N_B distinct χ's, the we really have a system of $(M + N_B)$ differential equations rather than M integro-differential equations. That is to say, the solution vector $Y(r) = [F(r), \chi(r)]$, that is, $(M + N_B)$ elements, can be solved non-iteratively by the method of Section 2.4.1.

The modified Bessel functions satisfy the following recursion relations (writing η for $\nu + \frac{1}{2}$)

$$I_{\eta-1}(\rho) - I_{\eta-1}(\rho) = \frac{2\eta}{\rho} I_\eta(\rho) \tag{2.201}$$

$$I_{\eta-1}(\rho) + I_{\eta+1}(\rho) = 2I_\eta(\rho)' \tag{2.202}$$

$$K_{\eta-1}(\rho) - K_{\eta+1}(\rho) = \frac{-2\eta}{\rho} K_\eta(\rho) \tag{2.203}$$

$$K_{\eta-1}(\rho) + K_{\eta+1}(\rho) = -2K_\eta(\rho)' \tag{2.204}$$

where the prime denotes differentiation with respect to ρ. We also have the relation

$$I_\eta(\rho)K_{\eta+1}(\rho) + I_{\eta+1}(\rho)K_\eta(\rho) = \frac{1}{\rho}. \tag{2.205}$$

We drop the suffices μ and ν in the subsequent discussion as they merely serve to qualify P and F, which are distinct symbols anyway; we shall write r for r_1. Let us define the first term in Eq. 2.199 by

$$T_\lambda^\alpha(pFr) \equiv \int_0^r dr_2\left(\frac{r}{r_2}\right)^{1/2} p(r_2)F(r_2)I_\eta(\alpha r_2)K_\eta(r). \tag{2.206}$$

Using Eq. 2.204, we have

$$\frac{d}{dr}\left(T_\lambda^\alpha \frac{1}{r^{1/2}K_\eta}\right) \rightarrow \frac{1}{r^{1/2}K_\eta}\frac{dT}{dr} - \frac{T}{2r^{3/2}K_\eta} + \frac{T\alpha}{2r^{1/2}K_\eta^{\,2}}\left(K_{\eta+1} + K_{\eta-1}\right)$$

$$= \frac{P(r)F(r)I_\eta}{r^{1/2}},$$

from Eq. 2.206, which can be rewritten as

$$\frac{dT}{dr} - \frac{T}{2r} = pFI_\eta K_\eta - \frac{T\alpha}{2K_\eta}(K_{\eta+1} + K_{\eta-1}). \tag{2.207}$$

We can rewrite our defining equation, (2.199), in the form

$$\frac{\chi_\lambda^\alpha}{r^{1/2}I_\eta} = \frac{T_\lambda^\alpha}{r^{1/2}I_\eta} + \int_r^\infty dr_2 p(r_2)\,\frac{F(r_2)K_\eta(\alpha r_2)}{r_2^{1/2}}$$

which we differentiate with respect to r and use Eq. 2.202 in obtaining the result

$$\frac{d\chi}{dr} - \frac{\chi}{2r} - \frac{\chi\alpha}{2I_\eta}[I_{\eta+1} + I_{\eta-1}] = \frac{dT}{dr} - \frac{T}{2r} - \frac{T\alpha}{2I_\eta}[I_{\eta+1} + I_{\eta-1}] - pFK_\eta I_\eta$$

Upon substituting for dT/dr from Eq. 2.207 we obtain

$$\frac{d\chi}{dr} = \frac{\chi}{2r} + \frac{\chi\alpha}{2I_\eta}(I_{\eta+1} + I_{\eta-1}) - \frac{T\alpha}{2}\left[\frac{I_{\eta+1} + I_{\eta-1}}{I_\eta} + \frac{K_{\eta+1} + K_{\eta-1}}{K_\eta}\right].$$

From Eq. 2.205 we can readily show that

$$K_\eta(I_{\eta+1} + I_{\eta-1}) + I_\eta(K_{\eta+1} + K_{\eta-1}) = \frac{2}{\alpha r}$$

hence

$$\frac{d\chi}{dr} = \frac{\chi}{2r} + \frac{\chi\alpha}{2I_\eta}(I_{\eta+1} + I_{\eta-1}) - \frac{T}{rI_\eta K_\eta}. \tag{2.208}$$

We now differentiate this result with respect to r, substituting for dT/dr from Eq. 2.207 and for T from Eq. 2.208 to obtain

$$\frac{d^2\chi_\lambda{}^\alpha(\mu\nu r)}{dr^2} = \chi_\lambda{}^\alpha\left[\alpha^2 + \frac{\lambda(\lambda+1)}{r^2}\right] - \frac{p_\mu(r)F_\nu(r)}{r}. \tag{2.209}$$

We have thus shown that we can replace the system of integro-differential equations, (2.196), by the system of differential equations given by (2.200) and (2.209). The χ-functions depend on the parameters λ, μ, and ν. In any practical problem we find that a particular $\chi_\gamma(\mu, \nu, r)$ may appear several times in the system of equations; in setting up the system (2.209) we only want distinct χ's, N_B of them say. We can now rewrite our system of coupled second-order ordinary differential equations as

$$\frac{d^2F_\mu}{dr^2} = \sum_{\nu=1}^{M} U_{\mu\nu}(r)F_\nu(r) + \sum_{i=1}^{N_B} \frac{g_i\chi_i(r)p_i(r)}{r}, \qquad \mu = 1, 2, \ldots, M$$

$$\frac{d^2\chi_i}{dr^2} = \left[\alpha^2 + \frac{\lambda_i(\lambda_i+1)}{r^2}\right]\chi_i - \frac{p_i(r)F_i(r)}{r}, \qquad i = 1, 2, \ldots, N_B. \tag{2.210}$$

The boundary conditions on F have been discussed previously; we now need to know the boundary conditions on χ. From Eq. 2.199 we have

$$\begin{aligned}\chi_\lambda^{(\alpha)}(r_1) &\underset{r_1\to 0}{\sim} r_1^{1/2} I_{\lambda+1/2}(\alpha r_1)\int_{r_1}^{\infty} dr p_\mu(r)\frac{F_\nu(r)K_{\lambda+1/2}(\alpha r)}{r^{1/2}} \\ &\underset{r_1\to 0}{\sim} E r_1^{1/2}(\alpha r_1)^{\lambda+1/2} \\ &\underset{r_1\to 0}{\sim} E_{\mu\nu\lambda}^{(\alpha)} r_1^{\lambda+1},\end{aligned} \tag{2.211}$$

where

$$E_{\mu\nu\lambda}^{(\alpha)} \equiv (\alpha)^{\lambda+1/2}\int_0^\infty dr r^{-1/2} p_\mu(r)F_\nu(r)K_{\lambda+1/2}(\alpha r), \tag{2.212}$$

and we have used the leading term in the series expansion of I about the origin. In the asymptotic domain, Eq. 2.199 gives

$$\begin{aligned}\chi_\lambda^{(\alpha)}(p_\mu F_\nu; r_1) &\underset{r_1\to\infty}{\sim} r_1^{1/2} K_{\lambda+1/2}(\alpha r_1) D_{\mu\nu\lambda}^{(\alpha)} \\ &\underset{r_1\to\infty}{\sim} r_1^{-\lambda} D_{\mu\nu\lambda}^{(\alpha)},\end{aligned} \tag{2.213}$$

where

$$D_{\mu\nu\lambda}^{(\alpha)} \equiv \int_0^\infty dr r^{-1/2} p_\mu(r)F_\nu(r)I_{\lambda+1/2}(\alpha r), \tag{2.214}$$

and we have used the leading term in the asymptotic series for K. For a fixed M, then, μ, ν, and λ will run over a finite number of distinct combinations (N_B in number) which we have subscripted collectively with i in Eq. 2.210.

In solving the system of equations (2.210), we must ensure that χ satisfies the boundary conditions (2.211) and (2.213), which cannot be achieved until we have solved our system itself for $F_\mu(r)$ for all values of r! One possible algorithm for solving the system is to extend the multiple solution method of Section 2.4.1. We now have $(M + N_B)$ second-order equations and so we need $2(M + N_B)$ integration constants. If $E_{\mu\gamma\lambda}$ are finite, then all the elements of the solution vector vanish at the origin, thereby determining $(M + N_B)$ integration constants. The remaining $(M + N_B)$ can be arrived at by generating $(M + N_B)$ linearly independent solutions out from the origin to the match point r_0 in the now familiar manner of setting one of the $(M + N_B)$ derivatives equal to unity and the remainder to zero and repeating the process $(M + N_B)$ times. In the region $r \geq r_0$ we again require $2(M + N_B)$ integration constants to identify the outer solution. From Eqs. 2.165 and 2.213 we know $(M - N_a) + N_B$ logarithmic derivatives, consequently there are $(M + N_a + N_B)$ unknown integration constants. We now generate $(M + N_a + N_B)$ linearly independent solutions in the outer region by setting to zero all of the A_μ, B_μ, C_τ, and D_k, except one, and repeating the process $(M + N_a + N_B)$ times.

The analog to the matching condition given in Eq. 2.240 is

$$\sum_{\alpha=1}^{M+N_B} u_\alpha \begin{bmatrix} \mathscr{F}_\mu \\ \mathscr{F}_\mu' \end{bmatrix}_{r_0}^{(\alpha)} = \sum_{\beta=1}^{M+N_a+N_B} w_\beta \begin{bmatrix} \mathscr{G}_\mu \\ \mathscr{G}_\mu' \end{bmatrix}_{r_0}^{(\beta)}, \tag{2.215}$$

which is a system of $2(M + N_B)$ equations in $2(M + N_B) + N_a$ unknowns, u and w. The remaining N_a equations are obtained in Section 2.4.1. This set of conditions guarantees that the channel functions $F_\mu^{(\nu)}(r)$ satisfy the boundary conditions at the origin, Eq. 2.167, and asymptotically, Eq. 2.163.

A numerical consistency check on the correctness of the solutions is provided by the boundary conditions on the χ-functions, see Eqs. 2.211 and 2.213. From the right-hand side of Eq. 2.215 we have

$$\begin{aligned} \chi_k(r_A) &= \sum_{\beta=1}^{M+N_a+N_B} w_\beta \mathscr{G}_k{}^\beta(r_A) \\ &= \sum_{\beta=M+N_a+1}^{M+N_a+N_B} w_\beta \mathscr{G}_k{}^\beta(r_A) \end{aligned}$$

since $\mathscr{G}_k{}^{\beta=1, M+N_a}(r_A) \equiv 0$ by construction. When the explicit form for $\mathscr{G}$

used to start the solution at r_A is substituted into this result we obtain

$$\chi_k(r_A) = \sum_{\beta=M+N_a+1}^{M+N_a+N_B} w_\beta \delta_{k\beta} r^{-\lambda(k)}, \qquad k = 1, 2, 3, \ldots, N_B$$
$$= w_k r^{-\lambda(k)},$$

hence

$$w_k^{(\alpha)} \equiv D_{\mu\nu\lambda}^{(\alpha)}.$$

We recall that we obtain N_a solutions to our matching equations, which we shall label j, hence

$$w_k^{(j)} = \int_0^\infty drr^{-1/2} p_\mu(r) F_\nu^{(j)}(r) I_{\lambda+1/2}(\alpha r). \tag{2.216}$$

From the left-hand side of Eq. 2.215 we have

$$\chi_k(\varepsilon) = \sum_{\alpha=1}^{M+N_B} u_\alpha \mathscr{F}_k^\alpha(\varepsilon)$$
$$= \sum_{\alpha=M+1}^{M+N_B} u_\alpha \delta_{k\alpha} \varepsilon^{\lambda(k)+1}$$
$$= u_k \varepsilon^{\lambda(k)+1},$$

which upon comparison with Eq. 2.211 leads to

$$u_k^{(j)} = \alpha^{\lambda+1/2} \int_0^\infty drr^{-1/2} p_\mu(r) F_\nu^{(j)}(r) K_{\lambda+1/2}(\alpha r). \tag{2.217}$$

The technique is to solve the matching equation N_a times for the matching parameters u and w and then evaluate the right-hand sides of Eqs. 2.216 and 2.217 to ensure the corresponding values of $w_k^{(j)}$ and $u_k^{(j)}$, $k = 1, 2, \ldots, N_B$ and $j = 1, 2, \ldots, N_a$ are obtained.

One of the practical problems in solving the equations is choosing r_A—the domain over which the $p_\mu(r)$, as well as the potentials, are tabulated. We can use Eqs. 2.216 and 2.217 as the criteria for choosing r_A, since in practice these integrals are not performed from 0 to ∞, but from 0 to r_A. We can increase r_A until the pair of self-consistency checks agree to within a specified epsilon.

Problem 29. In *L-S* coupling, the wave function describing the scattering of electrons by hydrogen-like ions is given by

$$\psi(\mathbf{r}_1, \mathbf{r}_2) = \tfrac{1}{2}(1 \pm P_{12}) \sum_{nl_1l_2} R_{nl_1}(r_1) F_{nl_1l_2}(r_2) \mathscr{Y}_{l_1l_2LM}(\hat{r}_1, \hat{r}_2) r_2^{-1}$$

where P_{12} interchanges the coordinates of the two electrons. Derive the form of the exchange term in the radial equations and show how its analog to χ of Eq. 2.199 satisfies a second-order differential equation [21].

2.4.4. Asymptotic Solutions

In Section 1.4.4 we considered the problem of generating a solution to a single second-order ordinary differential equation in the asymptotic domain. We now wish to generate linearly independent solutions $F_i(r)$ to the system of coupled equations given in (2.25), with $\eta_i \equiv (Z - N)/|k_i|$. Hence we shall present the analysis for the code used in ref. 25. We shall begin by making the fundamental assumption that we can expand the elements of the solution vector in the form

$$F_i(r) \underset{r\to\infty}{\sim} \sum_{\kappa=1}^{a}\left\{\sin\theta_\kappa \sum_{p=0}^{\infty} \alpha_p{}^{i\kappa} r^{-p} + \cos\theta_\kappa \sum_{p=0}^{\infty} \beta_p{}^{i\kappa} r^{-p}\right\} + \sum_{\tau=1}^{b} \exp\left(-|k_\tau|\, r + \eta_\tau \log 2\,|k_\tau|\, r\right) \sum_{p=0}^{\infty} \gamma_p{}^{i\tau} r^{-p}, \qquad (2.218)$$

where $a(b)$ is the number of distinct positive (negative) k_i^2. The r-independent coefficients, α, β, and γ, are to be found from recurrence relations. Of course, if any of the k_i^2 vanish, then Eq. 2.218 will not have a component representing such a solution; we should have to include the analogs of Eq. 1.208. At first sight we might expect the sums over κ and τ to be over the solutions of the N_a open and $N - N_a$ closed channels, respectively. However, such a sum would include redundant terms when

$$\theta_\kappa \equiv k_\kappa r + \eta_\kappa \log 2k_\kappa r. \qquad (2.219)$$

If we used the WKB method to generate functional forms for θ_κ which differed from channel to channel, then the sums over κ and τ would have to be extended over all channels. A method based on the WKB phases has been presented by Norcross and Seaton [82].

When Eq. 2.218 is substituted into Eq. 2.25 and the coefficients of $r^{-p}\sin\theta_\kappa$, $r^{-p}\cos\theta_\kappa$ and the exponential terms are set to zero independently, we are led to the three recurrence relations

$$(k_i^2 - k_\kappa^2)\alpha_p{}^{i\kappa} + [(p-2)(p-1) - \eta_\kappa{}^2]\alpha_{p-2}^{i\kappa} + 2k_\kappa(p-1)\beta_{p-1}^{i\kappa} + \eta_\kappa(2p-3)\beta_{p-2}^{i\kappa} = \sum_{j\lambda} a_{ij}{}^\lambda \alpha_{p-\lambda-1}^{j\kappa}, \qquad (2.220)$$

$$(k_i^2 - k_\kappa^2)\beta_p{}^{i\kappa} + [(p-2)(p-1) - \eta_\alpha{}^2]\beta_{p-2}^{i\kappa} - 2k_\kappa(p-1)\alpha_{p-1}^{i\kappa} - \eta_\kappa(2p-3)\alpha_{p-2}^{i\kappa} = \sum_{j\lambda} a_{ij}{}^\lambda \beta_{p-\lambda-1}^{j\kappa}, \qquad (2.221)$$

$$(k_i^2 + |k_\tau|^2)\gamma_p{}^{i\tau} + 2\,|k_\tau|\,(p-1)\gamma_{p-1}^{i\tau} + [\eta_\tau{}^2 + (p-2)(p-1)]\gamma_{p-2}^{i\tau} - (2p-3)\eta_\tau\gamma_{p-2}^{i\tau} = \sum_{j\lambda} a_{ij}{}^\lambda \gamma_{p-\lambda-1}^{j\tau}. \qquad (2.222)$$

We see that the third recurrence relation is decoupled from the first pair of relations; and that this latter pair has a potential source of error as $k_i^2 \to k_\kappa^2$, since this vanishing number will appear as a denominator in the formulas for computing higher α_p and β_p. In other words, we do not expect to get too close to thresholds with this algorithm for computing $F_i(r)$.

Since there are N equations, we shall need $2N$ integration constants. If $(N - N_a)$ of the elements of the solution vector vanish exponentially in the asymptotic region, then $N - N_a$ logarithmic derivatives are known and only $N + N_a$ of the integration constants are unknown. Thus we can generate $N + N_a$ linearly independent solutions

$$\{F_i(r)\}^\eta, \qquad \eta = 1, 2, \ldots, N + N_a,$$

by appropriate choices of α_0, β_0, and γ_0. All coefficients are set to zero except one; this nonzero coefficient is chosen in the following $N + N_a$ linearly independent ways

$$\begin{pmatrix} \alpha_0^{i\kappa(i)} \\ \beta_0^{i\kappa(i)} \\ \gamma_0^{i\tau(i)} \end{pmatrix} = \begin{pmatrix} \delta_{i\kappa(i)} \\ 0 \\ 0 \end{pmatrix}, \qquad i = 1, 2, \ldots, N_a;$$

$$\begin{pmatrix} \alpha_0^{i\kappa(i)} \\ \beta_0^{i\kappa(i)} \\ \gamma_0^{i\tau(i)} \end{pmatrix} = \begin{pmatrix} 0 \\ \delta_{i\kappa(i)} \\ 0 \end{pmatrix}, \qquad i = 1, 2, \ldots, N_a;$$

$$\begin{pmatrix} \alpha_0^{i\kappa(i)} \\ \beta_0^{i\kappa(i)} \\ \gamma_0^{i\tau(i)} \end{pmatrix} = \begin{pmatrix} 0 \\ 0 \\ \delta_{i\tau(i)} \end{pmatrix}, \qquad i = N_a + 1, \ldots, N.$$

In these relations, $\kappa(i)$ is that κ determined by i. We recall that in Eq. 2.218 κ only runs over the number of distinct $k_i^2 > 0$. For example, in e^-H scattering, including 1s-2s-2p, we shall have four channels, but only two distinct k_i^2 because of the 2s-2p degeneracy. If all channels are open, then

$$\kappa(1) = 1$$

$$\kappa(2) = 2$$

$$\kappa(3) = 2$$

$$\kappa(4) = 2.$$

2.5. CLOSE COUPLING APPROXIMATION

2.5.1. Electron-Atom

The importance of the radial equations we have been considering is due to their connection with variational principles for the parameters required in the calculation of cross sections, see Spruch [83]. To find the required variational principle, it is noted that the exact wave functions have the property that

$$\int \Psi^*(\Gamma, \mathbf{x}_1 \cdots \mathbf{x}_{N+1})(H - E)\Psi(\Gamma', \mathbf{x}_1 \cdots \mathbf{x}_{N+1})\, d\tau = 0. \quad (2.223)$$

The approximate calculation of the total wave functions is based on the following considerations.

Let us assume that we can construct approximate trial functions Ψ_τ which satisfy the true boundary conditions. In particular Ψ_τ has the desired form given by Eqs 2.12 and 2.163, although the elements of the **R**-matrix may be in error. Since the Hamiltonian is diagonal in $LM_LSM_S\Pi$, it is possible to consider the individual terms in Eq. 2.223, denoted

$$I_{\alpha\alpha'}^{LM_LSM_S\Pi} \equiv I_{\alpha\alpha'} = \int \Psi^*(\Gamma)(H - E)\Psi(\alpha' LM_LSM_S\Pi)\, d\tau = 0, \quad (2.224)$$

where α' is the channel index which we recall stands for all the other quantum numbers which make up the complete set. We consider small arbitrary variations of the wave functions of the type

$$\delta\Psi(\Gamma') \equiv \Psi_\tau - \Psi = (N+1)^{-\frac{1}{2}} \sum_{p=1}^{N+1} (-1)^{N+1-p} \sum_{\Gamma} \frac{\Psi(\Gamma, \mathbf{X}\hat{x}_p)\, \delta F_{\Gamma\Gamma'}(r_p)}{k_\Gamma r_p}, \quad (2.225)$$

with

$$\delta F_{\Gamma\Gamma'} \underset{r\to\infty}{\sim} k_\alpha^{-\frac{1}{2}}\, \delta R_{\alpha\alpha'} \cos\theta_{\alpha'},$$

about the true total wave function Ψ. For such variations, the first variation of I is given by

$$\delta I_{\alpha\alpha'} = \int \Psi^*(H - E)\, \delta\Psi\, d\tau + \int \delta\Psi^*(H - E)\Psi\, d\tau.$$

To reduce the first term on the right-hand side of this equation we use Green's theorem to obtain

$$\delta I_{\alpha\alpha'} = -\frac{1}{2}\int ds \left[F(\mathbf{r})\frac{\partial}{\partial n}\delta F(\mathbf{r}) - \delta F(\mathbf{r})\frac{\partial}{\partial n}F(\mathbf{r})\right] + \int \delta\Psi(H - E)\, \delta\Psi\, d\tau,$$

where n is normal to the surface s over which the integration is performed.

When Eq. 2.225 is used in this result, we obtain

$$\delta I_{\alpha\alpha'} = \frac{\delta R_{\alpha\alpha'}}{2} + \int \delta\Psi(H - E)\,\delta\Psi'\,d\tau.$$

When the integral term is neglected, because it is second order in small quantities, we have the Kohn variational principle

$$\delta\left(I_{\alpha\alpha'} - \frac{R_{\alpha\alpha'}}{2}\right) = 0. \tag{2.226}$$

When we write $\delta I = I_\tau - I$ and $\delta R = R_\tau - R$, then up to second order in the error in the wave function we have

$$R_{\alpha\alpha'} = R^\tau_{\alpha\alpha'} - 2I^\tau_{\alpha\alpha'}. \tag{2.227}$$

In the close-coupling approximation, the trial functions $F_\tau(r)$ are chosen such that $I^\tau = 0$. That is to say, the true values of the elements of the R-matrix are given by $\mathbf{R}^\tau$ up to $\delta\Psi'^2$.

The most general form for Ψ'_τ used to date is given in Eq. 2.142. When this form is substituted into I^τ we obtain

$$I_{kl}{}^\tau = \int\cdots\int d\mathbf{x}_1\cdots d\mathbf{x}_{N+1}\Bigg[(N+1)^{-1/2}\sum_{p=1}^{N+1}(-1)^{N+1-p}\sum_{\Gamma_i}\psi(\Gamma_i\mathbf{X}\hat{x}_p)F_{ik}(r_p)r_p{}^{-1}$$

$$+ \sum_\mu C_\mu{}^{\Gamma_k}\Phi_\mu(L_kS_k\Pi_k, \mathbf{x}_1\cdots\mathbf{x}_{N+1})\Bigg](H - E)\Psi'_\tau(\Gamma_l\mathbf{x}_1\cdots\mathbf{x}_{N+1}). \tag{2.228}$$

The step-by-step reduction of this expression is performed in Smith and Morgan [74]; only the result is of interest here, which allows Eq. 2.226 to be written as

$$\delta\Bigg[\sum_{ij}\int F_{ik}D_{ij}F_{jl}\,dr + \sum_{j\mu}C_\mu{}^k\int U_{\mu j}F_{jl}\,dr + \sum_{i\nu}C_\nu{}^l\int U_{\nu i}F_{ik}\,dr$$

$$+ \sum_{i\nu}C_\mu{}^kC_\nu{}^lA_{\mu\nu} - \tfrac{1}{2}R_{kl}\Bigg] = 0, \tag{2.228}$$

where

$$D_{ij} = -\frac{1}{2}\left[\frac{d^2}{dr^2} - \frac{l_i(l_i+1)}{r^2} + \frac{2Z}{r} + 2(E - \varepsilon_i)\right]\delta_{ij} + V_{ij} + W_{ij}.$$

The functions V, U, W, and A are all known; W being an integral operator.

For variations of F_{mn} of the form (2.225) we obtain

$$\sum_j D_{ij}F_{jl} + \sum_\mu C_\mu{}^lU_{\mu i} = 0, \tag{2.229}$$

while variations of (2.228) with respect to $C_\lambda{}^m$ lead to

$$\sum_{\nu} A_{\mu\nu} C_\nu{}^l + \sum_j \int U_{\mu j} F_{jl} \, dr = 0. \tag{2.230}$$

The numerical methods for solving these equations were presented in Section 2.4.

We now discuss a recent application [84] of Eq. 2.229 to illustrate the physical interpretation placed on the computed elements of R.

A detailed resonant structure in atomic lithium has been observed in photon absorption in the region 60–70 eV above the ground state by Ederer *et al.* [85]. Bound-state calculations of Weiss have served to identify the two lowest lying observed states as $(1s(2s2p)^3P)^2P$ and $(1s(2s2p)^1P)^2P$ and to identify the states immediately above this as a $(1s2s)^3S\ np$ series converging to the $(1s2s)^3S$ limit of Li^+. In order to clarify the resonant structure *above* the $(1s2s)^3S$ limit, close-coupling calculations involving the expansion of the total wave function in terms of the four $n = 2$ states of Li^+ $(1s2s^{1,3}S,$ $1s2p^{1,3}P)$ were performed in the region immediately above the 2^3S threshold. Specifically these results include the following:

1. Identification of the first 5 resonances above the 2^3S threshold.
2. Calculation of the quantum defects for all six series converging on the four series limits.
3. Tentative identification of the resonance observed at 193.58 Å which perturbs the $(1s, 2s)^3S\ np$ series.

These calculations were carried out with a computer code based on the analysis of Smith and Morgan [74]. In Fig. 29 we present a schematic energy level diagram which shows that for $L = 1$, there are three configurations involved in the e^--Li^+ problem. Because of computer storage limitations, calculations were carried out neglecting the distant ground state configuration of Li^+, namely $1s^2(^1S)$. In other words, the total wave function is approximated by expanding it as an antisymmetrized sum of products of bound-state wave functions corresponding to the $(1s2s)^{1,3}S$, $(1s2p)^{1,3}P$ terms of Li^+ and functions representing a bound or free electron in Li.

The bound-state wave functions used were those of Morse *et al.* [86]. The experimental values of the energies [87] corresponding to the $(1s2s)^{1,3}S$ and $(1s2p)^{1,3}P$ thresholds rather than the eigenvalues of the bound-state wave functions were used in the computations. Since Smith and Morgan chose to orthogonalize the orbitals describing the motion of the colliding electron with respect to the bound-electron orbitals, they included a virtual bound state, $(1s2s)^3S, 2p^2P$, constructed from the same orbitals used for the ion core states, to take into account that the colliding electron has a finite probability of spending some time in incomplete orbitals. This is the Φ term in Eq. 2.142.

Since this state is known to lie well below the $(1s2s)^3S$ ionization limit, [84] including it is expected to have little effect on the calculation. Since the ground state of lithium is ^2S (even), only ^2P (odd) states can be reached by optical absorption. Thus the e^-Li$^+$ calculations described here are limited to total quantum numbers $L = 1$ and $S = \frac{1}{2}$.

The output of the computer code consists of an R matrix which characterizes the asymptotic form of the wave functions of free electrons coupled to the ion core states and has dimensions corresponding to the number of open channels. In the present case (^2P (odd)), when all channels are open (above the 2^1P threshold) there are six open channels corresponding to free electron states $(1s2s)^{1,3}S$, εp, $(1s2p)^{1,3}P$, εs, and $(1s2p)^{1,3}P$, εd, respectively. In Fig. 29, these are the channels with $\pi = (-1)^L$, $S = \frac{1}{2}$. We define eigenphases (δ) via the matrix equation $\tan \delta = R$, where R is the diagonalized R matrix. The eigenphases above the 2^1P threshold as calculated are shown as continuous curves at the right of Fig. 30. These eigenphases provide *estimates* of the quantum defects for states converging in the four series limits via the relation $\pi\sigma_i = \delta_i$, where σ_i is called the quantum defect for each series and δ_i the eigenphase for each channel. Estimates of resonant positions have been obtained by extrapolating the calculated eigenphases below each threshold and are shown as triangles in Fig. 30. Observed positions of the first few members of the two observed series $[(1s2s)^3Snp]$ and $[(1s2p)^1Pns]$ and "effective" eigenphases, that is, $\pi\sigma$ observed, are shown as circles with the estimated experimental uncertainties indicated for the higher series members [85].

Direct calculations of the first five resonances above the 2^3S threshold were obtained by calculating the single eigenphase between the 2^3S and 2^1S thresholds and obtaining graphically the resonance position and width via the relation

$$\text{ctn}\,(\delta - \delta_0) = \frac{E - E_0}{\Gamma/2}.$$

These data are shown in Table 7, along with the observed resonant positions of corresponding absorption peaks from ref. 85. They are drawn as rectangles on Fig. 30. These computed data are preliminary, since the energy grid used in the calculations ($\sim$3 or 4 points for each resonance) was relatively coarse. The width of the first resonance is uncertain, since the background phase shift (δ_0) rises from $\sim$0.4 radians to $\sim$0.8 radians at higher energies. However, the energies of the four higher resonances are probably correct to 0.01 eV and their widths to the number of significant figures given in the table.

The eigenphase plot in Fig. 30 shows good agreement between the extrapolated eigenphases for the 2^1P *ns* series and the observed "effective"

Table 7. Resonances Above the 2^3S Threshold. Experimental Energies Represent Peaks in Absorption. The Theoretical Positions Were Obtained by Fitting the Phase Shift to the Expression ctn $(\delta - \delta_0) = (E - E_0)/1/2\Gamma$. Approximate Widths Obtained Using the Indicated Back ground Phase Shifts (δ_0) Appear in the Third Column. Resonant Energies Are Expressed As Energies Above the Ground State of Lithium

Theory	Experiment	Γ (eV)	δ_0	Classification
64.7	64.6–65.1	0.4	0.4	$(1s2s)\,^3S\,3p$
65.17	65.25	0.01	0.8	$(1s2p)\,^3P\,3s$
65.28	65.30	0.004	0.8	$(1s2p)\,^3P\,3d$
65.59	65.66	0.12	0.8	$(1s2p)\,^3P\,4s$
65.87	65.89	0.004	0.8	$(1s2p)\,^3P\,4d$

eigenphases, the calculated eigenphases being about 0.1 radians higher than those observed. The calculations also predict that the $2^1P\ nd$ series may be observable above the 2^3P limit and that it will have a small (~ -0.06) and negative quantum defect.

The resonant structure immediately above the 2^3S limit is complicated, since for lower series members the appropriate zero-order coupling is $(1s(2s\,np)^{1,3}P)^2P$ or $(1s(2p\,ns)^{1,3}P)^2P$ rather than $((1s2s)^{1,3}S\ np)^2P$ or $((1s2p)^{1,3}P\ ns)^2P$. The effect is to shift the energy levels for $n = 3$ states as indicated by the arrows in Fig. 30. This is confirmed by the direct calculations above the 2^3S threshold. The lowest calculated resonance lies above the expected position of $(1s2s)^1S\ \varepsilon p$ obtained by extrapolating the $2^1S\ \varepsilon p$ eigenphase. The resonance position is shown in Fig. 30 with an "effective" phase shift of $\sim$0.2. Similarly the $2^1P\ 3s$ and $2^3P\ 3d$ levels obtained from the extrapolation lie below the next two directly calculated levels. These levels, labeled as rectangles with integers 2 and 3 respectively inside, agree closely with the narrowly spaced doublet observed in absorption.

The agreement of the experimental and directly calculated positions for this doublet suggests that the broad region of absorption immediately above the 2^3S threshold is due to the first calculated resonance above threshold, $2^3P\ 3s$. It was conjectured by Cooper *et al.* that the perturbing level observed between the $n = 6$ and 7 levels of the $2^3S\ np$ series is the level $2^3P\ 3s$, which has been pushed below the 2^3S threshold by the strong interaction between the $n = 3$ states; see the upgoing arrow in Fig. 30. The "effective" phase shifts for the $2^3S\ np$ series and for the perturbing level (assuming it is bound to the 2^3P threshold) shown at the left of Fig. 30 confirm this suggestion.

The two highest resonances shown in Table 7, while they agree closely with observed peaks in absorption, require further investigation. A tentative identification of these resonances as $2^3P\ 4s$ and $2^3P\ 4d$ is indicated by the "effective" phase shifts shown in Fig. 30. Further analysis of the structure immediately below the 2^1S threshold will be difficult, since, due to the narrow spacing between the 2^1S and 2^3P thresholds, all three series converging to those limits will contribute to the resonant structure in that region.

In the preceding paragraphs we have merely introduced the reader to the problems of the analysis of phase shifts in the vicinity of resonances. Recently, there has been quite a lot of work [88] on resonance interpretation in Argand diagrams.

2.5.2. Atom-Atom

In a recent paper, Allison and Burke [89] have developed a theory of fine structure transitions in atom-atom collisions, using the basic techniques described in the preceding sections. This theory was based on several assumptions:

1. The first-order interaction is dominant.
2. Fine structure splitting can be neglected.
3. Initial and final states of the colliding systems must be represented by a single atomic energy term of a particular electronic configuration (namely, $1s^2 2s^2 2p^q$).

In this section we relax Assumption 3. That is to say, a theory is developed which allows the colliding neutral systems to be described by a linear superposition of different configurations employing distinct orbitals! Consequently, the theory presented will allow the computation of inelastic atom-atom cross sections involving the excitation of electronic configurations.

The antisymmetrized wave function describing an atomic system with N_a electrons has been written in Eq. 2.148; all quantities will have a subscript a.

A second atomic system with N_b electrons will be described by analogous expressions. We note explicitly that Eq. 2.148 as written, represents a single atomic term—configuration interaction has been neglected. We can construct the product wave function

$$|\gamma; \mathbf{X}_a\mathbf{X}_b\rangle \equiv [\phi(L_aS_aM_{L_a}M_{S_a}\Pi_a, \mathbf{X}_a) \times \phi(L_bS_bM_{L_b}M_{S_b}\Pi_b, \mathbf{X}_b)]^\gamma, \quad (2.231)$$

where $\times$ denotes the vector coupling of the N_a-electron function and the N_b-electron function, while γ denotes a complete set of quantum numbers, including

$$(S_aS_b)SM_S, (L_aL_b)L_tM_{L_t}.$$

We use these product wave functions as a basis for expanding the total wave function describing the slow collision between a pair of atomic systems, namely,

$$\Psi = \sum_\gamma |(S_aS_b)SM_S, (L_aL_b)L_tM_{Lt}\rangle\, |\gamma; \mathbf{R}\rangle$$

where the expansion coefficients can be written as

$$|\gamma; \mathbf{R}\rangle = \sum_{lm_l} |lm_l; \hat{R}\rangle\, |\gamma l m_l; R\rangle,$$

where $|lm_l; \hat{R})$ are simply the surface harmonics $Y_{lm_l}(\hat{R})$. When we combine these expressions vectorially we obtain

$$\Psi = \sum_{\gamma' l L M_L} |(S_a S_b) S M_S, ((L_a L_b) L_t l) L M_L\rangle \sum_{m_l M_{L_t}} (L_t l M_{L_t} m_l \mid L M_L)\, |\gamma l m_l; R\rangle,$$

where γ' denotes the complete set introduced above, less M_{L_t}. Let a new complete set of quantum numbers for the collision problem be defined by

$$\Gamma \equiv \gamma' l L M_L,$$

hence we shall rewrite this expression for Ψ as

$$\Psi \equiv \sum_{\Gamma} |\Gamma; a, b, \hat{R}\rangle \frac{F_\Gamma(R)}{R} \tag{2.232}$$

which defines the radial functions F_Γ. Initially, the system is in some particular state, Γ' say; hence we have

$$\Psi(\Gamma') = \sum_{\Gamma} |\Gamma; a, b; \hat{R}\rangle \frac{F_{\Gamma\Gamma'}(R)}{R}\,. \tag{2.233}$$

We note that Eq. 2.233 includes a sum over the atomic terms of a and b; consequently, from the asymptotic form of $F_{\Gamma\Gamma}$, we can extract the elements of the reactance matrix from which we can calculate the transition probabilities among the atomic terms of a and b.

The wave function, $\Psi(\Gamma)$, describing the relative motion of a and b, initially in the state Γ, satisfies the Schrödinger equation

$$\left[-\frac{\hbar^2}{2\mu}\nabla_R{}^2 + H_a + H_b + V\right]\Psi = E\Psi, \tag{2.234}$$

where H_a and H_b are the Hamiltonians for the isolated atomic systems, and μ is the reduced mass of the colliding systems:

$$\mu \equiv \frac{M_a M_b}{(M_a + M_b)}\,.$$

When we compute the matrix elements of this Hamiltonian, in the representation defined in Eq. 2.232, we see that the matrix is diagonal in LSM_LM_S. In other words, the radial functions need only be characterized by the channel indices, that is, $F_\nu(R)$, where

$$\nu \equiv S_a S_b,\ (L_a L_b) L_t l.$$

In order to derive the radial equations, we substitute (2.232) into (2.234), multiply throughout by $\langle\Gamma; a, b; \hat{R}|$, and integrate over $\mathbf{X}_a$, $\mathbf{X}_b$, and $\hat{R}$ to obtain

$$\left[\frac{d^2}{dR^2} - \frac{l_\nu(l_\nu + 1)}{R^2} + k_\nu^{\,2}\right]F_\nu(R) = \frac{2\mu}{\hbar^2}\sum_{\nu'}\langle\nu; a, b; \hat{R}|\, V\, |\nu'; a, b, \hat{R}\rangle\, F_{\nu'}(R) \quad (2.235)$$

where

$$k_\nu^{\,2} = \frac{2\mu}{\hbar^2}(E - E_a - E_b).$$

In this section we shall restrict our attention to the collision between two neutral, but possibly excited, configurations for which the first-order potential has been given by Rose [90] to be

$$\begin{aligned} V = \sum_{mn\alpha\alpha'\beta'} &(-1)^n(m + n)!\, R^{-m-n-1}\{(m - \alpha)!\,(m + \alpha)!\,(n - \alpha)!\,(n + \alpha)!\}^{-1/2} \\ &\times \sum_{l''}(mn\alpha'\beta' \mid l''\alpha' + \beta')(mn\alpha - \alpha \mid l''0)\left(\frac{4\pi}{2l'' + 1}\right)^{1/2} Y_{l''\alpha'+\beta'}(\hat{R})^* \\ &\times \left(\frac{4\pi}{2m + 1}\right)^{1/2}\sum_i e_i r_i^{\,m} Y_{m\alpha'}(\hat{r}_i)\left(\frac{4\pi}{2n + 1}\right)^{1/2}\sum_j e_j r_j^{\,n} Y_{n\beta'}(\hat{r}_j). \end{aligned} \quad (2.236)$$

When the explicit form of the wave functions is substituted into the matrix element on the right-hand side of Eqs. 2.235, the matrix element is given by

$$\begin{aligned} &\langle\nu; a, b, \hat{R}|\, V\, |\nu'; a, b, \hat{R}\rangle \\ &\quad = \langle(S_aS_b)SM_S, ((L_aL_b)L_tl)LM_L; \mathbf{X}_a\mathbf{X}_b\hat{R}|\, V\, |(S_{a'}S_{b'})SM_S, ((L_{a'}L_{b'})L_{t'}l')LM_L\rangle \\ &\quad = \sum (S_aS_bM_{S_a}M_{S_b} \mid SM_S)(L_aL_bM_{L_a}M_{L_b} \mid L_tM_{L_t})(L_tlM_{L_t}m_l \mid LM_L) \\ &\qquad \times (S_a'S_b'M_{S_a'}M_{S_b'} \mid SM_S)(L_a'L_b'M_{L_a'}M_{L_b'} \mid L_t'M_{L_t'})(L_t'l'M_{L_t'}m_{l'} \mid LM_L) \\ &\qquad \times \langle L_aS_aM_{L_a}M_{S_a}\Pi_a; \mathbf{X}_a|\sum_i e_i r_i^{\,m}\left(\frac{4\pi}{2m + 1}\right)^{1/2} \\ &\qquad \times Y_{m\alpha'}(\hat{r}_i)\,|L_a'S_a'M_{L_a'}M_{S_a'}\Pi_a'; \mathbf{X}_a\rangle \\ &\qquad \times \langle L_bS_bM_{L_b}M_{S_b}\Pi_b; \mathbf{X}_b|\sum_j e_j r_j^{\,n}\left(\frac{4\pi}{2n + 1}\right)^{1/2} \\ &\qquad \times Y_{n\beta'}(\hat{r}_j)\,|L_b'S_b'M_{L_b'}M_{S_b'}\Pi_b'; \mathbf{X}_b\rangle \\ &\qquad \times \int Y_{lm_l}(\hat{R})^*(-1)^n(m + n)!\, R^{-m-n-1} \\ &\qquad \times \{(m - \alpha)!\,(m + \alpha)!\,(n - \alpha)!\,(n + \alpha)!\}^{-1/2} \\ &\qquad \times (mn\alpha'\beta' \mid l''\alpha' + \beta')(mn\alpha - \alpha \mid l''0)\left(\frac{4\pi}{2l'' + 1}\right)^{1/2} \\ &\qquad \times Y_{l''\alpha'+\beta'}(\hat{R})^* Y_{l'm_{l'}}(\hat{R})\, d\hat{R}. \end{aligned} \quad (2.237)$$

In this equation we must evaluate an operator which is a linear superposition of one-electron operators, $r^m Y_{m\alpha'}(\hat{r})$, and which is symmetric under the interchange of the labels of any pair of electrons. Consequently, the matrix element can be written as

$$N_a \langle L_a S_a M_{L_a} M_{S_a} \Pi_a; \mathbf{X}_a| \, e r_{N_a}{}^m \left(\frac{4\pi}{2m+1}\right)^{1/2} Y_{m\alpha'}(\hat{r}_{N_a}) \, |L_a' S_a' M_{L_a'} M_{S_a'} \Pi_a'; \mathbf{X}_a\rangle$$

$$= eN_a[\mathscr{n}(N_\lambda{}^a)\mathscr{n}(N_\lambda{}^{a'})]^{-1/2} \sum_{q_a q_a'} (-1)^{P_{q_a}+P_{q_a'}} \langle q_a L_a S_a M_{L_a} M_{S_a} \Pi_a; \mathbf{X}_a|$$

$$\times \, r_{N_a}{}^m \left(\frac{4\pi}{2m+1}\right)^{1/2} Y_{m\alpha'}(\hat{r}_{N_a}) \, |q_a' L_a' S_a' M_{L_a'} M_{S_a'} \Pi_a'; \mathbf{X}_a\rangle. \tag{2.238}$$

The matrix element given in Eq. 2.238 vanishes if ϕ_a and ϕ_a', differ by more than one electron jump, when the electron orbitals associated with q_a are orthonormal to those of $q_{a'}$. We begin by assuming such orthonormality, as in Smith and Morgan [74]. Let ρ_a and ρ_a' be the subshells containing the interacting electron, label N_a, in ϕ_a and ϕ_a' respectively. Then the number of electrons in each subshell of ϕ_a and $\phi_{a'}$, namely $N_\lambda{}^a$ and $N_\lambda{}^{a'}$, are related by

$$\prod_\lambda \delta(N_\lambda{}^a, N_\lambda{}^{a'} + \delta_{\lambda\rho_a} - \delta_{\lambda\rho_a'}).$$

Furthermore, the nonzero contributions to the sums over q_a and $q_{a'}$ will come from those distributions which have the same distribution of spectator labels, that is, $\bar{q}$; let there be $\mathscr{n}(\bar{N}_\lambda)$ such distributions. Hence

$$N_a(\bar{N}_\lambda)[(N_\lambda{}^a)\mathscr{n}\,(N_\lambda{}^{a'})]^{-1/2} = \frac{N_a(N_a-1)!}{\prod_\lambda \bar{N}_\lambda} \left[\frac{\prod_\lambda N_\lambda{}^a!}{N_a!} \frac{\prod_\lambda N_\lambda{}^{a'}!}{N_a!}\right]^{1/2}$$

$$= [N_{\rho_a} N_{\rho_a'}]^{1/2}$$

When the above results are used in Eq. 2.238, and the interacting electron wave function is separated out, using the coefficients of fractional parentage, Eq. 2.238 reduces to

$$e[N_{\rho_a} N_{\rho_a'}]^{1/2} \sum_{\rho_a \rho_a'} \prod_\lambda \delta(N_\lambda{}^a, N_\lambda{}^{a'} + \delta_{\lambda\rho_a} - \delta_{\lambda\rho_a'})(-1)^{\Delta P}$$

$$\times \sum (l_{\rho_a}^{N\rho_a} \alpha_{\rho_a} S_{\rho_a} L_{\rho_a} \{| l_{\rho_a}^{\bar{N}\rho_a} \bar{\alpha}_{\rho_a} \bar{S}_\rho \bar{L}_{\rho_a}, l_{\rho_a})(l_{\rho_a'}^{\bar{N}\rho_a'} \bar{\alpha}_{\rho_a'} \bar{S}_{\rho_a'} \bar{L}_{\rho_a'}, l_{\rho_a'} |\} l_{\rho_a'}^{N\rho_a'} \alpha_{\rho_a'} S_{\rho_a'} L_{\rho_a'})$$

$$\times \int_0^\infty dr_{N_a} P_{\rho_a}(r_{N_a}) r_{N_a}{}^m P_{\rho_a'}(r_{N_a})$$

$$\times \langle \bar{L}_1 \bar{L}_2 \cdots [\bar{L}_{\rho_a} l_{\rho_a}(N_a)] L_{\rho_a} \cdots ; L_a| \left(\frac{4\pi}{2m+1}\right)^{1/2}$$

$$\times \, Y_{m\alpha'}(r_{N_a}) \, |\bar{L}_1 \bar{L}_2 \cdots [\bar{L}_{\rho_a'} l_{\rho_a'}(N_a)] L_{\rho_a'} \cdots ; L_{a'}\rangle$$

$$\times \langle \bar{S}_1 \bar{S}_2 \cdots [\bar{S}_{\rho_a} \tfrac{1}{2}(N_a)] S_{\rho_a} \cdots ; S_a \,|\, \bar{S}_1 \bar{S}_2 \cdots [\bar{S}_{\rho_a'} \tfrac{1}{2}(N_a)] S_{\rho_a'} \cdots ; S_{a'}\rangle \tag{2.239}$$

where

$$\Delta P \equiv \sum_{\lambda=\eta+1}^{\zeta} \bar{N}_\lambda,$$

and the spin recoupling coefficient is given by

$$\begin{aligned}
\delta_{S_aS_{a'}}\{\delta_{\eta\zeta} &+ (1-\delta_{\eta\zeta})[(2S_{\eta-1}^{\alpha'}+1) \\
&\times (2S_\eta+1)(2S_{\zeta-2}^{\alpha}+1)(2S_\zeta+1)]^{1/2}\}(-1)^{1/2+\bar{S}_\zeta-S_\zeta} \\
&\times W(S_{\eta-2}^{\alpha}\bar{S}_\eta S_{\eta-1}^{\alpha}\tfrac{1}{2};\, S_{\eta-1}^{\alpha'}S_\eta)W(S_{\zeta-2}^{\alpha'}\tfrac{1}{2}S_{\zeta-1}^{\alpha}\bar{S}_\zeta;\, S_{\zeta-2}^{\alpha}S_\zeta) \\
&\times [\delta_{\eta\zeta-1}+(1-\delta_{\eta\zeta-1})]\prod_i^{\zeta-3}[(2S_i^{\alpha}+1)(2S_{i+1}^{\alpha'}+1)]^{1/2} \\
&\times W(S_i^{\alpha}\bar{S}_{i+2}\tfrac{1}{2}S_{i+1}^{\alpha'};\, S_{i+1}^{\alpha}S_i^{\alpha'})
\end{aligned} \tag{2.240}$$

and where $\eta(\zeta)$ is the smaller (larger) of ρ_a and $\rho_{a'}$, $S^\alpha(S^{\alpha'})$ denote the intermediate coupling vectors in the state vector containing $\eta(\zeta)$, for example,

$$\{S_{\eta-2}^{\alpha}\bar{S}_\eta\}S_{\eta-1}^{\alpha}.$$

The Kronecker delta on S_a and $S_{a'}$ ensures that the sum over the spin magnetic quantum numbers in Eq. 2.237 is unity.

To evaluate the orbital matrix element, under the orthonormality constraint, we shall modify slightly the technique devised by Fano *et al.* [52] for two-electron operators. We recall that this involved introducing a fictitious particle, the orbiton. We rewrite the one-electron operator in the form

$$\left(\frac{4\pi}{2m+1}\right)^{1/2} Y_{m\alpha'}(\hat{r}_{N_a}) \equiv \sum_{\bar{\alpha}} C_{\bar{\alpha}}^{\ m}\langle m\bar{\alpha} \mid m\alpha'\rangle.$$

From Fano *et al.* we have

$$\begin{aligned}
&\langle\{\bar{L}_1\bar{L}_2\cdots[\bar{L}_\eta l_\eta]L_\eta\}L_{\eta-1}^{\alpha}\cdots;\, L^\eta|\sum_{\bar{\alpha}} C_{\bar{\alpha}}^{\ m}\,\langle m\bar{\alpha}| \\
&\qquad = \sum_{\bar{l}}(2l_\eta+1)^{-1/2}(l_\eta\|\, C^m\,\|\bar{l})\,\langle\bar{L}_1\bar{L}_2\cdots[\bar{L}_\eta,(\bar{l}m)l_\eta]L_\eta\cdots\bar{L}_\zeta\cdots;\, L^\eta|,
\end{aligned}$$

where the reduced matrix element is given by

$$(l_\eta\|\, C^m\,\|\bar{l}) = (2\bar{l}+1)^{1/2}(\bar{l}m\,00\mid l_\eta 0) \tag{2.241}$$

and $L^\eta(L^\zeta)$ is $L_a(L_{a'})$ if $\rho_a(\rho_{a'})$ is the inner (outer) of ρ_a and $\rho_{a'}$. The vector addition of the orbiton state vector, $|m\alpha'\rangle$, and the ket is given by

$$\begin{aligned}
&|m\alpha'\rangle\,|\bar{L}_1\bar{L}_2\cdots\bar{L}_\eta\cdots[\bar{L}_\zeta l_\zeta]L_\zeta\cdots;\, L^\zeta M_{L^\zeta}\rangle \\
&\qquad = \sum_{\bar{L}M_{\bar{L}}}(L^\zeta m M_{L^\zeta}\alpha'\mid \bar{L}M_{\bar{L}})\,|\bar{L}_1\bar{L}_2\cdots\bar{L}_\eta\cdots[\bar{L}_\zeta l_\zeta]L_\zeta\cdots,\, L^\zeta m;\, \bar{L}M_{\bar{L}}\rangle
\end{aligned}$$

These results are combined to give the orbital matrix element in the form

$$\langle \bar{L}_1 \bar{L}_2 \cdots [\bar{L}_{\rho_a} l_{\rho_a}] L_{\rho_a} \cdots ; L_a M_{L_a} | \left(\frac{4\pi}{2m+1}\right)^{1/2}$$

$$\times\ Y_{m\alpha'}(\hat{r})\, |\bar{L}_1 \bar{L}_2 \cdots [\bar{L}_{\rho_{a'}} l_{\rho_{a'}}] L_{\rho_{a'}} \cdots ; L_{a'} M_{L_{a'}}\rangle$$

$$= \left[\frac{(2l_\zeta + 1)}{(2l_\eta + 1)}\right]^{1/2} (l_\zeta m 00 \mid l_\eta 0)(L^\zeta m M_{L\zeta} \alpha' \mid L^\eta M_{L^\eta})$$

$$\times\ \langle \bar{L}_1 \bar{L}_2 \cdots [\bar{L}_\eta, (l_\zeta m) l_\eta] L_\eta \cdots \bar{L}_\zeta \cdots ; L^\eta \mid$$

$$\times\ \bar{L}_1 \bar{L}_2 \cdots \bar{L}_\eta \cdots [\bar{L}_\zeta l_\zeta] L_\zeta \cdots L^\zeta m; L^\eta\rangle \qquad (2.242)$$

It is possible to write out explicitly, in terms of Racah coefficients, the orbital recoupling coefficient given in Eq. 2.242. For example, when $\rho_a = \rho_{a'} =$ the outermost subshell, as in Allison and Burke, then

$$\langle [\bar{L}_{\rho_a}, (l_{\rho_a} m) l_{\rho_a}] L_a | [[\bar{L}_{\rho_a} l_{\rho_a}] L_{a'}, m] L_a$$

$$= [(2L_{a'} + 1)(2l_{\rho_a} + 1)]^{1/2} W(\bar{L}_{\rho_a} l_{\rho_a} L_a m; L_{a'} l_{\rho_a}). \quad (2.243)$$

In evaluating the matrix elements in Eq. 2.237 we need not make the assumption that the electron orbitals associated with the atomic term a are orthonormal to those associated with a'. In other words, to each term in the atomic system a we can assign a distinct set of electron orbitals. Consequently, it will be possible to take full advantage of the atomic structure models which include configuration interaction. For example, Froese-Fischer [91] imposes the orthogonality conditions

$$\int_0^\infty P(nl; r) P(n'l; r)\, dr = \delta_{nn'},$$

only on orbitals within a configuration.

In evaluating the matrix element appearing in Eq. 2.237,

$$\left\langle L_a S_a M_{L_a} M_{S_a} \Pi_a; \mathbf{X}_a \right| \sum_i e_i r_i^{\,m} \left(\frac{4\pi}{2m+1}\right)^{1/2} Y_{m\alpha'}(\hat{r}_i) \left| L_a' S_a' M_{L_a'} M_{S_a'} \Pi_a' ; \mathbf{X}_a \right\rangle$$

we note that the operator is symmetric under the interchange of any pair of labels, while the state vectors are antisymmetric; consequently, each member of the distributions making up the bra will contribute equally; hence the matrix element becomes

$$\left[\frac{\mathscr{N}(N_\lambda)}{\mathscr{N}(N_{\lambda'})}\right]^{1/2} \sum_q (-1)^{P_q} \langle q_0 L_a S_a M_{L_a} M_{S_a} \Pi_a; \mathbf{X}_a | \sum_{i=1}^{N_a} e_i r_i^{\,m} \left(\frac{4\pi}{2m+1}\right)^{1/2}$$

$$\times\ Y_{m\alpha'}(\hat{r}_i)\, |q L_a' S_a' M_{L_a'} M_{S_a'} \Pi_a'; \mathbf{X}_a\rangle, \quad (2.244)$$

where the aggregate $q_0 = \{q_\lambda^0\}$ represents the normal distribution of the N_a

labels in groups of N_λ. We now proceed in precisely the same way as in evaluating Eq. 2.149, leading to the definition of *a* G-function analogous to that given in Eq. 2.151.

In order to simplify the expression for the potential, Eq. 2.237, we perform the integration over $\hat{R}$,

$$\int Y^*_{lm_l} Y^*_{l''\alpha'+\beta'} Y_{l'm_{l'}}\, d\hat{R} = \left[\frac{(2l+1)(2l''+1)}{4\pi(2l'+1)}\right]^{1/2} (ll''00 \mid l'0)(ll''m_l\alpha'+\beta' \mid l'm_{l'}),$$

and substitute both the analog to Eq. 2.151 and the similar expression for system b into Eq. 2.237 to give

$$\begin{aligned}
&\langle \nu; a, b, \hat{R}|\, V\, |\nu'; a, b, \hat{R}\rangle \\
&= \sum (S_aS_bM_{S_a}M_{S_b} \mid SM_S)(S_aS_bM_{S_a}M_{S_b} \mid SM_S)(L_tlM_{L_t}m_l \mid LM_L) \\
&\quad \times (L_aL_bM_{L_a}M_{L_b} \mid L_tM_{L_t})(L_a'L_b'M_{L_a'}M_{L_b'} \mid L_{t'}M_{L_{t'}})(L_t'l'M_{L_{t'}}m_{l'} \mid LM_L) \\
&\quad \times \left[\frac{(2l+1)(2l''+1)}{4\pi(2l'+1)}\right]^{1/2} (ll''00 \mid l'0)(ll''m_l\alpha'+\beta' \mid l'm_{l'}) \left(\frac{4\pi}{2l''+1}\right)^{1/2} \\
&\quad \times (L_amM_{L_a'}\alpha'' \mid L_a\, M_{L_a})(L_b'nM_{L_b'}\beta' \mid L_b\, M_{L_b})(mn\alpha'\beta' \mid l''\alpha'+\beta') \\
&\quad \times (mn\alpha-\alpha \mid l''0)(-1)^n(m+n)!\, R^{-m-n-1} \\
&\quad \times [(m-\alpha)!\,(m+\alpha)!\,(n-\alpha)!\,(n+\alpha)!]^{-1/2} \\
&\quad \times G(L_aS_a\Pi_a; m; L_a'S_a'\Pi_a')G(L_bS_b\Pi_b; n; L_b'S_b'\Pi_b'). \qquad (2.245)
\end{aligned}$$

The sum over the spin magnetic quantum numbers gives unity; the sum over the orbital magnetic quantum numbers can be performed, using the technique of Section 2.2, and the result is

$$\begin{aligned}
&\langle \nu; a, b, \hat{R}|\, V\, |\nu'; a, b, \hat{R}\rangle \\
&= \sum_{mn} (-1)^{n+L+l+L_t+\sigma} R^{-m-n-1}(m+n)! \\
&\quad \times [(2L_a+1)(2L_b+1)(2l+1)(2l'+1)(2L_t+1)(2L_{t'}+1)]^{1/2} \frac{(2L+1)}{(2l''+1)} \\
&\quad \times G(L_aS_a\Pi_a; m; L_a'S_a\Pi_a')G(L_bS_b\Pi_b; n; L_b'S_b\Pi_b') \\
&\quad \times \sum_\alpha [(m-\alpha)!\,(m+\alpha)!\,(n-\alpha)!\,(n+\alpha)!]^{-1/2} \sum_{l''} (mn\alpha-\alpha \mid l''0) \\
&\quad \times (ll'00 \mid l''0) W(L_t'l'L_tl; Ll'') \begin{pmatrix} L_{t'} & L_{a'} & L_{b'} \\ L_t & L_a & L_b \\ l'' & m & n \end{pmatrix}, \qquad (2.246)
\end{aligned}$$

where the index σ denotes the sum of the nine elements of the Wigner $9j$ coefficient.

Problem 30. Show that Eq. 2.246 reduces to Eq. 10 of Allison and Burke.

The system of equations given in Eq. 2.235 can be solved by the methods described in Section 2.4 and the resulting R-matrix elements extracted and used in the total cross-section formula

$$\sigma(E) = \frac{\pi}{k^2} \sum \frac{(2L + 1)(2S + 1)}{(2L_a + 1)(2L_b + 1)(2S_a + 1)(2S_b + 1)} |T^{LS}_{L_t'l', L_tl}|^2 \quad (2.247)$$

where T is defined in Eq. 1.42. In Fig. 31 we present the total cross section for $O(^3P) - O(^3P)$ scattering as computed by Allison and Burke, using this theory.

2.5.3. Scattering by Diatomic Molecules

In this final section we consider the scattering of a structureless projectile by a rigid rotator within the close-coupling approximation. That is to say, the diatomic molecule is assumed to be equivalent to a rigid rotator with moment of inertia, I. One of the most significant papers on this subject is by Arthurs and Dalgarno [92], who took into account in their formulation the coupling between different energy levels of the rotator.

Let $\hat{r}' = \theta'\phi'$ specify the orientation of the internuclear axis with respect to axes fixed in space. From wave mechanics we know that the eigenfunction of the target rotator is $Y_{l_1m_1}(\hat{r}')$, where $l_1\hbar$ is the orbital angular momentum of the rotator, and is a solution of

$$H_{\text{rot}} Y_{l_1m_1}(\hat{r}') = \frac{\hbar^2}{2I} l_1(l_1 + 1) Y_{l_1m_1}(\hat{r}'). \quad (2.248)$$

Let $\hat{r} = \theta\phi$ specify the direction of motion of the incident particle at a distance r from the center of mass of the rotator. The Schrödinger equation describing the overall system is

$$\left[H_{\text{rot}} - \frac{\hbar^2}{2\mu} \nabla_r^{\,2} + V(r, \theta - \theta')\right] \psi(l_1 l_2 LM) = E\psi, \quad (2.249)$$

where μ is the reduced mass of the projectile in the center-of-mass frame, $V(r, \theta - \theta')$ is the interaction potential between the colliding systems, and $l_2\hbar$ is the orbital angular momentum of the projectile. The total orbital angular momentum vector of the complete system of projectile plus target is

$$\mathbf{L} = \mathbf{l}_1 + \mathbf{l}_2. \tag{2.250}$$

The radial motion of the projectile can be separated from the angular variables by the expansion

$$\psi(l_1 l_2 LM; \mathbf{r}\hat{r}') = \sum_{l_1' l_2'} Y_{l_1' l_2' LM}(\hat{r}, \hat{r}') \frac{F^L_{l_1' l_2', l_1 l_2}(r)}{r}, \tag{2.251}$$

where, according to Eq. 2.61,

$$Y_{l_1 l_2 LM}(\hat{r}, \hat{r}') = \sum_{m_1 m_2} (l_1 l_2 m_1 m_2 \mid LM) Y_{l_1 m_1}(\hat{r}) Y_{l_1 m_2}(\hat{r}').$$

Equation 2.251 is substituted into Eq. 2.249, Eq. 2.248 is used to eliminate H_{rot}, the result is premultiplied by Y and integrated over all the angular coordinates, to yield the radial motion of the projectile relative to the center-of-mass of the rotator

$$\left[\frac{d^2}{dr^2} - \frac{l_2'(l_2'+1)}{r^2} + k^2_{l_1'}\right] F^L_{l_1' l_2', l_1 l_2}(r) + \sum_{l_1'' l_2''} \langle l_1'' l_2'' LM | V | l_1' l_2' LM \rangle F^L_{l_1'' l_2'', l_1 l_2}(r) = 0, \tag{2.252}$$

where

$$\langle l_1'' l_2'' LM | V | l_1' l_2' LM \rangle = \frac{2}{\hbar^2} \iint d\hat{r}\, d\hat{r}'\, Y_{l_1'' l_2'' LM}(\hat{r})^* V Y_{l_1' l_2' LM},$$

and

$$k^2_{l_1'} = \frac{2m}{\hbar^2}\left[E - \frac{\hbar^2}{2I} l_1'(l_1'+1)\right]$$

$$= k_{l_1}{}^2 + \frac{m}{I}[l_1(l_1+1) - l_1'(l_1'+1)].$$

The problem of scattering by a rigid rotator has been reduced to the solution of coupled systems of second-order ordinary differential equations, namely Eq. 2.252.

Up to this stage we have not discussed the form of the potential V. Arthurs and Dalgarno proposed the expansion

$$V(r, \theta - \theta') = \sum_{\mu} v_\mu(r) P_\mu(\hat{r}, \hat{r}'), \qquad \mu = 0, 1, 2, \ldots \tag{2.253}$$

which upon substitution into the matrix element we have

$$\langle |V| \rangle = \sum_{\mu} v_\mu(r)(-1)^{l_1'+l_2'-L}[(2l_1'+1)(2l_2'+1)(2l_1''+1)(2l_2''+1)]^{1/2} \times (2\mu+1)^{-1}(l_2' l_2'' 00 \mid \mu 0)(l' l'' 00 \mid \mu 0) W(l_1' l_2' l_1'' l_2''; L\mu). \tag{2.254}$$

Problem 31. Prove Eq. 2.254 is true.

From the properties of the Clebsch-Gordan coefficients, with zero magnetic quantum numbers, we have that

$$l_1' + l_1'' + \mu = \text{even number},$$

and

$$l_2' + l_2'' + \mu = \text{even number}.$$

From these selection rules we have that

$$(l_1' + l_2') + (l_1'' + l_2'') \text{ is even.}$$

In other words, if initially $l_1' + l_2'$ is odd (even), then in the final state $l_1'' + l_2''$ is odd (even). This is a statement of the conservation of parity, namely,

$$(-1)^{l_1'+l_2'} = (-1)^{l_1''+l_2''}.$$

Equation 2.254 determines the number of coupled channels, N say, of which N_a only may be open, leading to an $N_a \times N_a$ R-matrix, from which we can calculate the cross section

$$\sigma(l_1 \to l_1') = \frac{\pi}{(2l_1 + 1)k_{l_1}^2} \sum_{L=0}^{\infty} \sum_{l_2=|L-l_1|}^{L+l_1} \sum_{l_2'=|L-l_1'|}^{L+l_1'} (2L + 1)\,|T^L(l_1'l_2', l_1l_2)|^2 \tag{2.255}$$

where we recall that T is expressed in terms of the R-matrix by

$$\mathbf{T} \equiv -2i\mathbf{R}(\mathbf{1} - i\mathbf{R}).$$

The detailed computational procedure for the scattering of atoms by diatomic molecules has been presented by Lester and Bernstein [93] and is an example of the general procedures presented in Section 2.4. Precisely the same techniques have been used by Lane and Geltman [94] for the electron-diatomic molecule case. However, the above formalism does not take electron exchange into account. The formalism for including exchange effects in the rotational excitation of molecular hydrogen by slow electrons has been given by Ardill and Davison [95] and used by Henry and Lane [96]. As expected, these effects convert the system of Eq. 2.251 to integro-differential equations.

Henry [97] has extended the preceding analysis to include vibrational as well as rotational excitation in electron-molecular hydrogen scattering.

References

[1] R. M. Thaler, *Lectures in Theoretical Physics*, **IV**, Boulder, Colo., 1961; Interscience 1962, p. 393.
[2] L. I. Schiff, *Quantum Mechanics*, 2nd ed., McGraw-Hill, 1955, p. 96.
[3] N. F. Mott and H. S. W. Massey, *Theory of Atomic Collisions*, 3rd ed., Oxford University Press, 1965, p. 21.
[4] L. D. Landau and E. M. Lifshitz, *Quantum Mechanics*, 2nd ed., Pergamon, Sec. 35 1958, and 3rd ed., 1965.
[5] R. G. Newton, *J. Math. Phys.* **1**, 319 (1960); *Scattering Theory of Waves and Particles*, McGraw-Hill, 1966.
[6] J. M. Blatt and V. F. Weisskopf, *Theoretical Nuclear Physics*, Wiley, 1952, p. 68.
[7] A. F. J. Siegert, *Phys. Rev.*, **56,** 750 (1939).
[8] U. Fano, *Phys. Rev.* **124,** 1866 (1961).
[9] U. Fano and J. W. Cooper, *Phys. Rev.* **137,** A1364 (1965).
[10] E. Gerjuoy, *Autoionization*, A. Temkin, Ed., Mono Book Corp. 1966, p. 33.
[11] K. W. Ford and J. A. Wheeler, *Ann. Phys.* **7,** 287, 259 (1959). See also R. G. Newton, *Scattering Theory of Waves and Particles*, McGraw-Hill, 1966.
[12] B. L. Moiseiwitsch and S. J. Smith, *Rev. Mod. Phys.* **40,** 238, 1968, present a comprehensive review of analytical approximation schemes which will not be considered at all here.
[13] N. F. Mott and I. N. Sneddon, *Wave Mechanics and Its Applications*, Oxford, 1948, p. 156. Also Bush and Caldwell, *Phys. Rev.* **38,** 1898 (1931).
[14] R. B. Bernstein, *J. Chem. Phys.* **33,** 795, (1960).
[15] H. S. W. Massey and R. A. Buckingham, *Proc. Roy. Soc.* **A168,** 378 (1938).
[16] R. A. Buckingham, J. Hamilton, and H. S. W. Massey, *Proc. Roy. Soc.* **A179,** 103 (1941).
[17] R. A. Buckingham, A. R. Davies, and D. G. Gillies, *Proc. Phys. Soc.* **71,** 457 (1958).
[18] M. J. Seaton and G. Peach, *Proc. Phys. Soc.* **79,** 1296 (1962).
[19] H. B. Dwight, *Tables of Integrals and Other Mathematical Data*, 4th ed., Macmillan Co., 1961, p. 76.
[20] A. Burgess, *Proc. Phys. Soc.* **81,** 442 (1963).
[21] D. R. Hartree, *The Calculation of Atomic Structures*, John Wiley and Sons, Inc., 1957, p. 71.
[22] I. H. Sloan, *J. Comp. Phys.* **2,** 414 (1968).
[23] E. C. Ridley, *Proc. Cambridge Phil. Soc.* **51,** 702 (1955).
[24] P. G. Burke and H. M. Schey, *Phys. Rev.* **126,** 147 (1962).
[25] P. G. Burke, D. D. McVicar, and K. Smith, *Proc. Phys. Soc.* **83,** 397 (1964).
[26] E. L. Ince, *Ordinary Differential Equations*, Dover, New York, 1944, p. 540.
[27] S. Chandrasekhar and F. H. Breen, *Astrophys. J.* **103,** 41 (1946).
[28] G. V. Marr, *Photoionization Processes in Gases*, Academic Press, 1967, Section 2.8.
[29] N. A. Doughty, "The Negative Hydrogen Ion Absorption Coefficients," Ph.D. Dissertation, University of W. Ontario, 1964.
[30] S. Chandrasekhar, *Astrophys. J.* **102,** 223 (1945).
[31] N. A. Doughty and P. A. Fraser, *Monthly Notices Roy. Astron. Soc.*, 132, 267 (1966).
[32] J. M. Blatt and L. C. Biedenharn, *Rev. Mod. Phys.* **24,** 258 (1952).
[33] E. H. S. Burhop, *Quantum Theory*, *I.*, Academic Press, 1961, p. 369.

[34] J. M. Blatt and V. F. Weisskopf, *Theoretical Nuclear Physics*, John Wiley and Sons, Inc., 1952, pp. 521–530.
[35] K. Smith, *Rept. Prog. Phys.* **29,** II, p. 373 (1966).
[36] U. Fano, *Phys. Rev.* **124,** 1866 (1961).
[37] P. K. Carroll, R. E. Huffman, J. C. Larrabee, and Y. Tanaka, *Astrophys. J.* **146,** 553 (1966).
[38] K. Smith and S. Ormonde, *Phys. Rev. Lett.* **25,** 563 (1970).
[39] R. J. Eden and J. R. Taylor, *Phys. Rev.* **133,** B1575 (1964).
[40] M. Gailitis and R. Damburg, *Proc. Phys. Soc.* **82,** 192 (1963).
[41] M. J. Seaton, *Proc. Phys. Soc.* **77,** 174 (1961).
[42] L. Fonda, *Ann. Phys.* **29,** 401 (1964).
[43] R. J. W. Henry, P. G. Burke, and A.-L. SinFaiLam, *Phys. Rev.* **178,** 218 (1969).
[44] J. Hunt and B. L. Moiseiwitsch, *J. Phys. B. Atom. Molec. Phys.* **3,** 892 (1970).
[45] H. A. Bethe, *Phys. Rev.* **76,** 38 (1949).
[46] A. R. Edmonds, *Angular Momentum in Quantum Mechanics*, Princeton University Press, 1957.
[47] M. E. Rose, *Elementary Theory of Angular Momentum*, John Wiley and Sons, Inc., 1957.
[48] M. Rotenberg, R. Bivins, N. Metropolis, and J. K. Wooten, Jr., *The 3j and 6j Symbols*, M.I.T. Press, 1959.
[49] E. U. Condon and G. S. Shortley, *Theory of Atomic Spectra*, Camb. Univ. Press, 1935, Chapter III, pp. 45–54.
[50] L. C. Biedenharn, J. M. Blatt, and M. E. Rose, *Rev. Mod. Phys.* **24,** 249 (1952).
[51] G. Racah, *Phys. Rev.* **62,** 438 (1942).
[52] U. Fano, F. Prats, and Z. Goldschmidt, *Phys. Rev.* **129,** 2643 (1963).
[53] U. Fano, *Phys. Rev.* **140,** A67 (1965).
[54] K. Smith and L. A. Morgan, *Phys. Rev.* **165,** 110 (1968).
[55] I. C. Percival and M. J. Seaton, *Proc. Cambridge Phil. Soc.* **53,** 654 (1957).
[56] L. C. Biedenharn, *J. Math. Phys.* **31,** 287 (1952).
[57] A. Arima, H. Horie, and Y. Tanabe, *Prog. Theor. Phys.* **11,** 143 (1954).
[58] H. A. Jahn and J. Hope, *Phys. Rev.* **93,** 318 (1954).
[59] R. J. Ord-Smith, *Phys. Rev.* **94,** 1227 (1954).
[60] P. G. Burke, *Comp. Phys. Comm.* **1,** 241 (1970).
[61] G. Racah, *Phys. Rev.* **63,** 367 (1943).
[62] T. Ishidzu and S. Obi, *Phys. Soc.* (*Japan*) **5,** 142 (1950).
[63] H. A. Jahn, *Phys. Rev.* **96,** 989 (1954).
[64] M. J. Englefield, *Phys. Rev.* **98,** 1213 (1955).
[65] A. M. Lane and R. G. Thomas, *Rev. Mod. Phys.* **30,** 2911 (1958).
[66] U. Fano, *Rev. Mod. Phys.* **29,** 74 (1957).
[67] R. Hagedorn, unpublished report, CERN 58-7 (1958).
[68] A. Simon, *Phys. Rev.* **92,** 1050 (1953).
[69] P. G. Burke and H. M. Schey, *Phys. Rev.* **126,** 163, 1962.
[70] A. Simon and T. A. Welton, *Phys. Rev.* **90,** 1036 (1953).
[71] K. Smith and M. Peshkin, *Argonne National Laboratory Report*, ANL-5910 (1959).
[72] R. J. W. Henry and L. Lipsky, *Phys. Rev.* **153,** 51 (1967).
[73] L. A. Morgan, Ph.D. Dissertation, London University, 1968.
[74] K. Smith and L. A. Morgan, *Phys. Rev.* **165,** 110 (1968).
[75] Helium: R. P. Madden and K. Codling, *Astrophys J.* **141,** 364, (1965); Nitrogen: P. K. Carroll, R. E. Huffman, J. C. Larabee, and Y. Tanaka, *Astrophys. J.* **146,** 553, (1966); Neon: K. Codling, R. P. Madden, and D. L. Ederer, *Phys. Rev.* **155,** 26,

(1967); Argon: H. E. Levy, and R. E. Huffman, J. Quant. *Spectrosc. Radiative Transfer* (*GB*), **9,** No. 10, pp. 1349–1358 (October, 1969).; Sulphur: G. Tondello, Astrophys J. (1971).

[76] R. J. W. Henry, *Phys. Rev.* **162,** 56 (1967).

[77] P. G. Burke, *Comp. Phys. Comm.* **1,** 241 (1970).

[78] K. Smith and P. G. Burke, *Phys. Rev.* **123,** 174 (1961).

[79] B. H. Bransden and J. S. C. McKee, *Proc. Phys. Soc.* **A67,** 422 (1956).

[80] K. Smith, R. J. W. Henry, and P. G. Burke, *Phys. Rev.* **147,** 21 (1966).

[81] N. F. Mott and I. N. Sneddon, *Wave Mechanics and Its Applications*, Oxford Clarendon Press, 1948.

[82] D. Norcross and M. J. Seaton, *J. Phys. B.* (*Atom. Mol. Phys.*) **2,** 731 (1969).

[83] L. Spruch, *Lectures in Theoretical Physics XI-C*, S. Geltman *et al.*, Ed., Gordon & Breach, p. 77, 1969.

[84] J. W. Cooper, M. J. Conneely, K. Smith, and S. Ormonde, *Phys. Rev. Lett.* **25,** 1540, (1970).

[85] D. L. Ederer, T. Lucatorto, and R. P. Madden, *Phys. Rev. Lett.* **25,** 1537, (1970).

[86] P. M. Morse, L. A. Young, and E. S. Haurwitz, *Phys. Rev.* **48,** 948 (1935).

[87] C. E. Moore, *Atomic Energy Levels*," *NBS Circular 467*, U.S. Govt. Printing Office, Wash., D.C., 1948. Note that the $1s2s\,^1S$ level of Li^+ (490079 cm^{-1} above the ground state) has recently been found to be 491373 cm^{-1}. See C. H. Pekeris, *Phys. Rev.* **126,** 143 (1962), and the corrected value has been used in the calculation reported here.

[88] F. Halzen and P. Minkowski, *Nucl. Phys.* **A14,** 522 (1969); J. Macek and P. G. Burke, *Proc. Phys. Soc.* **92,** 351 (1968).

[89] D. C. S. Allison and P. G. Burke, *J. Phys. B.* (*Atom Mol. Phys.*) **2,** 941 (1969).

[90] M. E. Rose, *J. Math. Phys.* (M.I.T.) **37,** 215 (1958).

[91] C. Froese-Fischer, *Comp. Phys. Comm.* **1,** 151 (1970).

[92] A. M. Arthurs and A. Dalgarno, *Proc. Roy. Soc.* **A256,** 540 (1960).

[93] W. A. Lester, Jr., and R. B. Bernstein, *J. Chem. Phys.* **48,** 4896 (1968).

[94] N. F. Lane and S. Geltman, *Phys. Rev.* **160,** 53 (1967).

[95] R. W. B. Ardill and W. D. Davison, *Proc. Roy. Soc.* **A304,** 465 (1968).

[96] R. J. W. Henry and N. F. Lane, *Phys. Rev.* **183,** 221 (1969).

[97] R. J. W. Henry, *Phys. Rev.* **A2,** 1349 (1970).

Index